TRANSFORM TECHNIQUES FOR PROBABILITY MODELING

OPERATIONS RESEARCH AND INDUSTRIAL ENGINEERING

Consulting Editor: J. William Schmidt

CBM, Inc., Cleveland, Ohio

Applied Statistical Methods
I. W. Burr

Mathematical Foundations of Management Science and Systems Analysis
J. William Schmidt

Urban Systems Models
Walter Helly

Introduction to Discrete Linear Controls: Theory and Application
Albert B. Bishop

Integer Programming: Theory, Applications, and Computations
Hamdy A. Taha

Transform Techniques for Probability Modeling
Walter C. Giffin

In preparation:

Analysis of Queueing Systems
J. A. White, J. W. Schmidt, and G. K. Bennett

TRANSFORM TECHNIQUES FOR PROBABILITY MODELING

WALTER C. GIFFIN

Department of Industrial and Systems Engineering
The Ohio State University
Columbus, Ohio

ACADEMIC PRESS New York San Francisco London 1975

A Subsidiary of Harcourt Brace Jovanovich, Publishers

ACADEMIC PRESS, INC.
111 Fifth Avenue, New York, New York 10003

United Kingdom Edition published by
ACADEMIC PRESS, INC. (LONDON) LTD.
24/28 Oval Road, London NW1

Library of Congress Cataloging in Publication Data

Giffin, Walter C (date)
Transform techniques for probability modeling.

(Operations research and industrial engineering)
Bibliography: p.
Includes index.
1. Integral transforms. 2. Transformations (Mathematics) 3. System analysis. 4. Probabilities. I. Title.
QA432.G54 519.2 74-17990
ISBN 0-12-282750-3

PRINTED IN THE UNITED STATES OF AMERICA

To
Beverly, Steven, and Becky

CONTENTS

Chapter 5 TRANSFORMS AND RANDOM EFFECTS IN LARGER SYSTEMS

Chapter 6 APPLICATIONS FROM THE THEORY OF QUEUES

Chapter 7 LINEAR SYSTEMS ANALYSIS

PREFACE

This book is concerned with a collection of mathematical techniques centered about various transforms which the author believes are especially applicable to systems analysis and probability modeling. The viewpoint adopted is that of the engineer. Rigorous mathematical proofs are often bypassed for more intuitively appealing if less precise arguments. The approach is pragmatic, implying an operating philosophy of "if it works use it."

The topics discussed were selected on the basis of three observations made by this author as he assembled course material for undergraduate and graduate level courses in Industrial and Systems Engineering at The Ohio State University.

(1) Transform techniques have wide applicability in many areas of operations research and systems analysis but are rarely discussed in a unified context-oriented manner in the literature.

(2) Transfer of knowledge from one area of study to another is often inhibited by special terminology introduced in newer disciplines which may obscure well-known relationships from older more established fields. For example, discussions of characteristic functions in many elementary texts in probability and statistics are rarely accompanied by discussions relating those functions to the much broader area of Fourier transforms.

(3) Operations from the calculus of finite differences, although widely applicable to the discrete functions common in the study of economic systems, are rarely understood by most undergraduate students.

Taken as individual areas of study, every topic treated herein has been discussed in much greater depth in other more specialized textbooks. The intended contribution of this work is to make available to fledgling systems analysts, operations researchers, management scientists, and industrial engineers the most important operational tools from each of those areas, without intimidating them with a great many academic credit hours of advanced math courses. At the same time it is hoped that the manner of presentation is such that those practitioners with prior exposure to some or all of the topics may be led to new understanding of their tools by tracing the strong connections among them.

Most of the material has been used in a combined advanced undergraduate and graduate level course in mathematical methods for industrial engineers taught by the author at Ohio State. The preparation of those students included basic engineering calculus and some exposure to elementary probability and statistics.

ACKNOWLEDGMENTS

Appreciation is hereby expressed to the many students who helped polish lectures by asking incisive questions and suffering impossible exams. Thanks also to my colleagues in the Department of Industrial and Systems Engineering at The Ohio State University who provided valuable feedback and suggestions stemming from the interface between my course and theirs. Indeed, most of the text content was originally designed to meet specific needs of more advanced modeling courses in that department.

My special thanks goes to my former department chairman, Dr. William T. Morris. An outstanding scholar and personal friend, Dr. Morris provides the best example possible for younger colleagues to emulate. In addition to inspiration and encouragement, he has given that most precious of academic commodities, the time to think and to carry out the mechanics of preparing a manuscript. The secretarial staff of Joan Case, Dixie Green, and Carol McDonald deserve special thanks for the difficult task of reducing my sometimes illegible scrawl to a more readable typed copy.

My intellectual debt obviously includes countless named and unnamed teachers and authors who molded my outlook over the years. They deserve any credit this work may reflect upon their contribution to my education. Any defects in the work belong to the author.

Finally, let me thank my wife, Beverly, and children, Steven and Rebecca. Their faith in the ultimate success of this venture is perhaps the most important of all.

CHAPTER 1

INTRODUCTION

The core subject matter of this book concerns the application of various transform techniques to problems in systems analysis and probability modeling. In the process of developing some facility with these techniques, including Fourier, Laplace, exponential, geometric, and z-transforms, auxiliary topics augmenting the main thrust will be discussed. Among such topics are selected tools from linear algebra, differential equations, finite calculus, and Fourier series. The goal is to create a mathematical montage which will hopefully provide new skills and appreciations for fledgling systems analysts, especially those concerned with nondeterministic systems.

The Pragmatic Approach

The view adopted in this book is that the validity of a concept is tested by its practical results. We will not be concerned with the inherent beauty of abstract mathematical manipulations but rather with the real-world decision problems and system understanding which can be aided by judicious application of certain classes of mathematical techniques.

There are many real-world problems which defy solution through formal mathematical analysis if one demands perfection in terms of exact models. The best model may be one which is a gross approximation if viewed in terms of total system fidelity but which is easily manipulated, easily understood, and offers insight into selected aspects of system structure. The prime demand we will place upon a mathematical technique is that it be useful in contributing to our understanding of system behavior.

Students of science and engineering have long been accustomed to employing approximations to analyze and design systems. Few studies involving empirical data are accomplished without attempting at some point in the analysis to approximate empirical curves with a judiciously selected mathematical expression. Even those studies involving very well-defined mathematical expressions may require approximations in order to carry out required calculations. For example, the use of the first few terms of the Taylor series expansion for complex functions encountered in elementary calculus is an approximation often used to obtain a more easily manipulated expression.

Many of the techniques discussed in this book are presented in a similar vein. The complete mathematical model may be unwieldy or even impossible to obtain. In such cases, we will employ Taylor series, Fourier series, the method of moments, or any other viable technique which may give a reasonable approximation to the true system behavior. We will be operating in the best engineering tradition. If a technique seems to work, use it.

Finite Calculus

The calculus of finite differences is a topic which is at first glance peripheral to the main thrust of this book. The main reason for devoting an extensive appendix to such a discussion is to provide the reader with some special-purpose tools which will be especially useful in calculations involving infinite sums of discrete functions. It is the experience of this author that computational skills relating to discrete functions have not been well developed by most undergraduate and graduate students in engineering. Our technique will be to show the close parallels between sums and integrals and between differences and derivatives. The resulting identities will later be used to solve difference equations and evaluate discrete transforms.

Transform Genealogy

Our entree to the field of operational calculus with its many variations of transforms will be by way of the Fourier series representation of a function. The Fourier series representation of periodic functions is introduced first as a means of obtaining a trigonometric approximation for unwieldy or discontinuous functions untreatable by Taylor's series. Applications in curve fitting and forecasting illustrate the direct engineering usefulness of the Fourier series.

When nonperiodic functions are considered, it is a short move in logic to develop the Fourier integral. Representation of that integral by means of the so-called Fourier transform pair will then lead us directly into the study of operational calculus techniques.

Since calculations performed using any of the common transforms are closely parallel as we move from transform to transform, we will concentrate much of our technique development on Laplace transforms. The Laplace transform is introduced as a special case of the Fourier transform. The reason for emphasizing Laplace transforms is that extensive tables are available for use by the analyst. Our approach will be that of the engineer who will use existing identities wherever possible in preference to evaluating complex inversion formulas. Such a view is greatly facilitated by the availability of extensive tables of Laplace transform pairs.

In addition to the Fourier and Laplace transforms, the closely related exponential transform will be introduced, with their connections duly noted. The z-transform and the geometric transform will be introduced as discrete versions of the Laplace and exponential transforms. Finally, the Mellin transform will be discussed as it relates to problems involving products and quotients of random variables.

Basic identities for each of the transforms discussed will be developed. Their use in reducing the computational burden in routine problems involving derivatives, integrals, differences, and sums will be illustrated. However, the big payoff for time spent developing these basic manipulative skills comes with the discussions of probability implications.

Probability Modeling

In many basic courses in probability and statistics students are introduced to several forms of "generating functions." They are shown characteristic functions, moment generating functions, and probability generating functions with little or no comment relating these to their more proper names of Fourier, exponential, and geometric transforms. Our first goal will be to forge that link and in so doing open the very broad field of operational calculus to fledgling system analysts who otherwise might never be able to solve any but the most routine problems involving "generating functions."

While on the subject of probability modeling, we will emphasize the techniques available for finding moments of compound functions of random variables. For instance, we will extensively explore sums of random variables, and random sums of random variables. Such exploration will be accompanied by many illustrations of real-world problem contexts in which such combinations arise.

Illustrative Examples

In order to demonstrate the applicability of the various transform techniques to probability modeling topics currently appearing in the literature, discussions of Markov chains and queueing theory will be included. The subject of linear system analysis is also included to offer the reader a brief glimpse of another major area in which transform techniques have proven extremely useful. This topic is treated from the point of view of a designer of management control systems. The concepts of convolution operations, transfer functions, and feedback are introduced in such contexts as inventory systems and forecasting models. This is somewhat different from the usual electrical engineering format used by many authors. However, the goal is essentially the same. The intent is to provide the reader with a set of mathematical tools to facilitate the analysis and design of often complex physical systems, which in our case may contain human elements. These discussions demonstrate how a basic knowledge of transforms, and an introduction to their special probability implications, can be used to analyze fairly sophisticated stochastic processes. Enough theory concerning these processes is presented to enable those with no prior exposure to understand the applications of the transforms techniques.

Problem Exercises

As is true of most mathematical modeling skills, one develops such capabilities only by repeated attempts to apply the basic principles to which he has been exposed. For that reason, the problem sets at the end of each chapter should be considered by the reader as an integral part of the presentation. These exercises range from simple applications of various definitions through reasonably complex model-building problems. Many of the advanced problems require the reader to extrapolate from the basic information in the text discussion. The idea is that by forcing one to extrapolate and synthesize, rather than merely memorize, a higher level of understanding can be achieved.

CHAPTER 2

BASICS OF OPERATIONAL CALCULUS

The purpose of this chapter is to introduce the reader to the basic techniques of operational calculus. The treatment which follows is somewhat "cookbook" in nature. The reason for presenting this material is to make the book reasonably self-contained for those readers with little or no prior exposure to operational calculus. Those readers with prior knowledge of transform techniques should skip this chapter and proceed directly to Chapter 3.

Operational Methods

In a classical sense the term "operational method" originally implied a procedure for solving differential equations where transformations involving algebraic operators for differentiating and integration were used. The newer concept of the operational method is, however, based on a functional transformation such as those provided by Fourier and Laplace integrals.

One could think of the act of differentiation as a transformation. In early courses in calculus, functions $f(t)$ are transformed to functions $f'(t)$ by the operator D. That is,

$$D(f(t)) = f'(t)$$

The function $f'(t)$ is the transform of $f(t)$ under the transformation. For example, we could view the function $2t$ as the transform of t^2.

In like fashion the act of integration can also be viewed as a transformation. The operation

$$I(f(t)) = \int_0^x f(t)\,dt$$

transforms the function $f(t)$ to a functional $f(x)$. Even the operation of multiplying all functions by the same constant can be viewed as a transformation.

In each of the familiar cases just cited, inverse transformations exist. We know that when a transform is given, a function $f(t)$ exists that has that transform, subject to certain limitations. The transform $D(f(t))$, for example, applies only to all differentiable functions. The point is that the idea of transforms and inverse transforms is nothing new to anyone who has studied ordinary algebra and calculus. What we seek now are techniques to enhance the use of such operational methods in increasingly complex problems.

All of the transformations used in this book will be linear. This means that if $f_1(t)$ and $f_2(t)$ are functions to be transformed by the operator T, then for any arbitrary constants c_1 and c_2 it is true that

$$T(c_1 f_1(t) + c_2 f_2(t)) = c_1 T(f_1(t)) + c_2 T(f_2(t)) \tag{2-1}$$

The particular transforms we will discuss are the Fourier, Laplace, exponential, Mellin, geometric, and z-transforms.

The integral transforms are often first encountered in a course in differential equations. Transforms are extremely useful in solving ordinary differential equations. In solving an nth-order linear differential equation, for example, one can often use Laplace transforms to reduce the equation to a simple algebraic expression in the transform of the unknown function $f(t)$. At the same time the initial or boundary conditions are automatically incorporated in the transform solution. This contrasts with classical methods, in which the general solution must be fitted to these conditions. Solutions are often expedited through the use of tables and a small set of theorems about the characteristics of the transform. The transform method can also provide a great amount of intuition about the physical system being modeled.

In addition to their more familiar uses in solving differential and difference equations, transforms can also be helpful in solving certain classes of integral equations and integrodifferential equations. In particular, the convolution property of transforms can be used to reduce a sometimes messy integration task for finding the distributions of combinations of random variables into simple algebraic operations in the transform domain. These techniques constitute the major thrust of this book.

In every instance, our use of transforms will be motivated by a desire to replace formidable mathematical calculations involving differentiation and integration with much simpler algebraic functions in transforms. In a sense, transforms give us capabilities in calculus much as logarithms gave us capabilities in simple algebra. For example, the product of two numbers can be replaced by a simple addition of logarithms to ease the burden of computation. In a similar manner the convolution of two functions can be replaced by a simple product of transforms. In both cases, one makes use of tables in moving from the transform to its inverse and back.

Relationships among Transforms

One interesting method for visualizing interrelationships among the transforms used in this book is to trace each of them back to the Fourier series representation of a function.[1]

Fourier Series

In elementary calculus, one often employs the Taylor series representation of a function. That is,

$$f(t) = \sum_{k=0}^{\infty} \frac{f^k(a)}{k!}(t - a)^k \tag{2-2}$$

where the function and its derivatives are evaluated at $t = a$. A common ploy is to take the first few terms of this expansion as an approximation for the function $f(t)$ when one finds a power series a more convenient form with which to work. The expansion requires that $f(t)$ and its derivatives be continuous at the point $t = a$.

When one encounters discontinuous functions, a more convenient expansion to use is the Fourier series. The Fourier series represents a function as a sum of sine and cosine terms of increasing frequency. The amplitude of these terms is determined by integration rather than differentiation as required for the Taylor series. For that reason, discontinuities can be tolerated.

Initially it is helpful to consider only periodic functions. A function $f(t)$ is said to be periodic if there exists a constant $2p$ such that

$$f(t + 2p) = f(t) \qquad \text{for all} \quad t \tag{2-3}$$

[1] For additional detail on this approach see Wylie [35, pp. 285–295].

The most familiar periodic functions are the ordinary trigonometric functions. The function sin x, for example, is periodic with period 2π. In simple terms that means that the function repeats itself every 2π units on the x-axis. It is apparent that this function could also be viewed as being periodic with period 4π, 6π, 8π, etc. However, for our purposes the term "period" will be preserved for the smallest number, $2p$, that satisfies Eq. (2-3). This is sometimes referred to as the primative period.

Periodic functions with forms other than simple sines and cosines are often encountered in mechanical as well as economic systems. For example, in machining operations a bad bearing on a lathe may introduce a periodic disturbance causing vibration problems and attendant surface irregularities on the product. In forecasting sales at the neighborhood bar, a two-week pay period at a nearby factory may elicit periodic disturbances in the demand pattern for beer. In both cases the precise form of the periodic function may be discontinuous, have a variety of shapes over limited lengths of time, and in general be very awkward to handle in a model. Our purpose in using the Fourier series is to approximate such phenomena with easily manipulated functions.

Beginning with a periodic function of period $2p$, the Fourier series representation is given by

$$f(t) = a_0 + \sum_{n=1}^{\infty} a_n \cos\frac{n\pi t}{p} + b_n \sin\frac{n\pi t}{p} \tag{2-4}$$

where

$$a_0 = \frac{1}{2p}\int_d^{d+2p} f(t)\,dt$$

$$a_n = \frac{1}{p}\int_d^{d+2p} f(t)\cos\frac{n\pi t}{p}\,dt, \qquad n > 0$$

$$b_n = \frac{1}{p}\int_d^{d+2p} f(t)\sin\frac{n\pi t}{p}\,dt, \qquad n > 0$$

and d is any arbitrary point on the t axis. As one takes more and more terms of the summation on the right, the composite curve becomes as close as desired to the function $f(t)$.

To illustrate how quickly the Fourier series approximates discontinuous functions, consider a paint-spray booth which operates on a 10-minute cycle. It requires five minutes to unload and reload the booth. The paint is then sprayed continuously for five minutes at a constant rate of three gallons per minute to complete the cycle. In order to model the master control system for a shop having many such booths, it is desirable to obtain a continuous approximation for paint demand at each booth.

Defining the origin as a point in time at which spraying begins, the function to be approximated is

$$f(t) = \begin{cases} 3, & 0 < t < 5 \\ 0, & 5 < t < 10 \end{cases} \qquad \text{where} \quad f(t+10) = f(t)$$

Integrating from Eq. (2-4), we find that

$$a_0 = \frac{1}{10}\int_0^5 3\,dt = 1.5$$

$$a_n = \frac{1}{5}\int_0^5 3\cos\frac{n\pi t}{5}\,dt = 0, \qquad n = 1, 2, \ldots$$

$$b_n = \frac{1}{5}\int_0^5 3\sin\frac{n\pi t}{5}\,dt = \frac{3(1-\cos n\pi)}{n\pi}, \qquad n \geq 1$$

from which

$$f(t) = 1.5 + \sum_{n=1}^{\infty} \frac{3(1-\cos n\pi)}{n\pi}\sin\frac{n\pi t}{5}$$

or

$$f(t) = 1.5 + \frac{6}{\pi}\left(\sin\frac{\pi t}{5} + \frac{1}{3}\sin\frac{3\pi t}{5} + \frac{1}{5}\sin\frac{5\pi t}{5} + \cdots\right) \tag{2-5}$$

The effect of including more terms of the Fourier series in the approximation is graphically illustrated in Figure 2-1. Even with only the first three

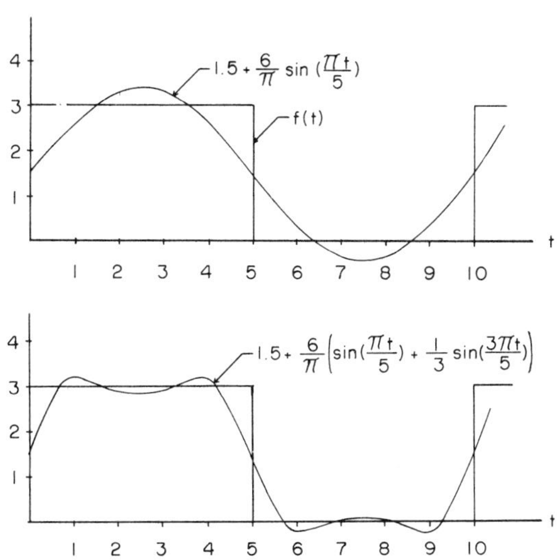

Figure 2-1 Fourier series approximations.

nonzero terms of the series, the approximation to this discontinuous function is very close. The reader is invited to plot additional terms of the series to see their effect.

Nonperiodic Functions

Since most functions encountered in this book are not periodic, it is desirable to extend such a series representation to nonperiodic functions. One way to do this for a function defined over a finite interval is to treat the function as though it had a periodic extension. If, for example, the function is

$$f(t) = \begin{cases} t, & 0 \le t \le 5 \\ 0, & \text{elsewhere} \end{cases}$$

one might choose to view this as one segment of a much larger periodic function. The segment from zero to five from any one of the periodic functions of Figure 2-2 could be used as $f(t)$. Since we are only concerned with that small interval, acting as though the function continued on in some periodic fashion will not compromise the accuracy of the series approxima-

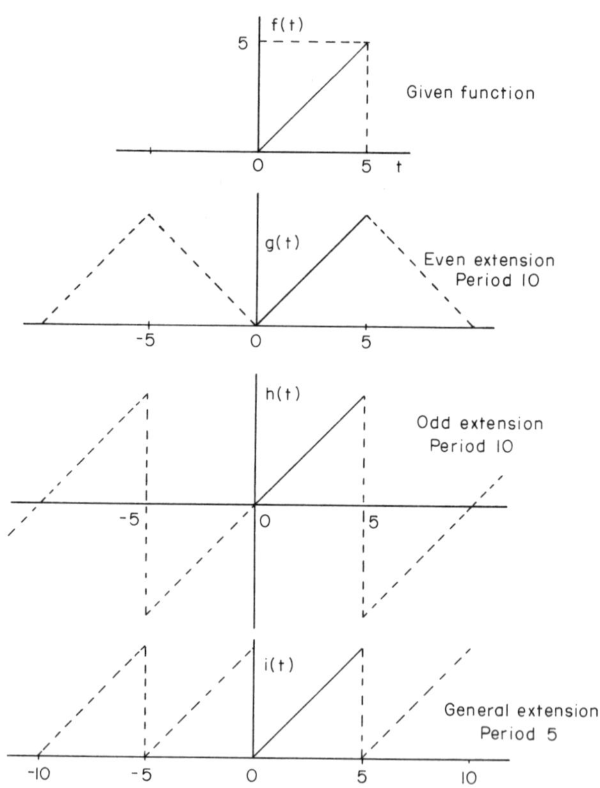

Figure 2-2 Periodic extensions.

tion in that interval. As long as the function is bounded and piecewise continuous, the limit of the series will converge to $f(t)$ where $f(t)$ is continuous, and to the average value of the right- and left-hand limits of $f(t)$ at each point where $f(t)$ is discontinuous [35, p. 248].

In each extension illustrated, the function is assumed to be periodic. When the extension is assumed to be even or odd, one has the basis for the so-called half-range expansions. A function is even if

$$f(-t) = f(t) \tag{2-6}$$

and it is odd if

$$f(-t) = -f(t) \tag{2-7}$$

The advantage of using an even or odd extension over just any periodic extension is that computation is reduced as many of the coefficients go to zero. For even functions of period $2p$

$$a_n = \frac{2}{p}\int_0^p f(t)\cos\frac{n\pi t}{p}\,dt, \qquad b_n = 0 \tag{2-8}$$

and for odd functions

$$a_n = 0, \qquad b_n = \frac{2}{p}\int_0^p f(t)\sin\frac{n\pi t}{p}\,dt \tag{2-9}$$

Fourier Transforms

To extend these ideas to nonperiodic functions, which may be defined for the entire real line instead of for a finite interval, it is convenient to rewrite the series in an equivalent exponential form. From trigonometry

$$\sin x = \frac{e^{ix} - e^{-ix}}{2i} \tag{2-10}$$

$$\cos x = \frac{e^{ix} + e^{-ix}}{2} \tag{2-11}$$

where $i^2 = -1$. Using these identities, we can write the Fourier series as

$$f(t) = \sum_{n=-\infty}^{\infty} c_n \exp\frac{in\pi t}{p} \tag{2-12}$$

where

$$c_n = \frac{1}{2p}\int_d^{d+2p} f(t)\exp\frac{-in\pi t}{p}\,dt \tag{2-13}$$

Consider some arbitrary function $f(t)$ defined in the interval $[-a, a]$ and again treat it as a periodic function. This time treat the period as some value $2p \geq 2a$. Figure 2-3 illustrates how this function might appear as the period

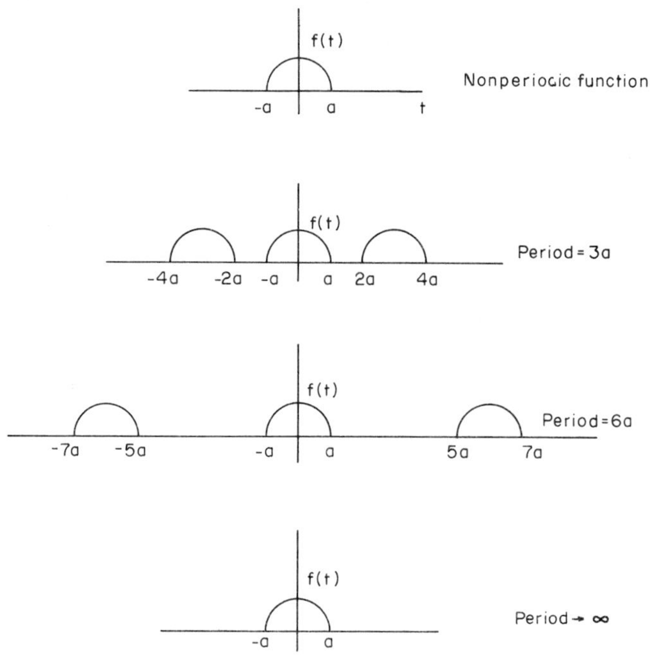

Figure 2-3 The limit of a periodic function.

becomes larger. Obviously if we allow the period to go to infinity we are back with the original nonperiodic function $f(t)$. This then suggests that we might start with the Fourier series representation of $f(t)$ as though $f(t)$ has some arbitrary period $2p$ and then take the limit as p goes to infinity.

The frequency of such a periodic process is most often viewed as the number of cycles completed by that process per unit time, where the period is the time required to complete a single cycle. In manipulating the common trigonometric functions, however, it is convenient to think of frequency in terms of radians per second. This means that sin $(5t)$ can be said to have a frequency of five radians per second or equivalently $5/2\pi$ cycles per second.

Let us define the frequency of the nth term in Eq. (2-12) as $w_n = \pi n/p$ radians per unit time and subscript the function $f(t)$ to denote its dependence upon some arbitrary period $2p$. After substituting (2-13) into (2-12) and multiplying by π/π, the exponential form of the Fourier series representation of $f(t)$ becomes

$$f_p(t) = \sum_{n=-\infty}^{\infty} \left[\frac{1}{2p}\int_d^{d+2p} f_p(t) \exp(-iw_n t)\, dt\right] \exp(iw_n t)$$

$$= \frac{1}{2\pi}\sum_{n=-\infty}^{\infty} \left[\int_d^{d+2p} f_p(t) \exp(-iw_n t)\, dt\right][\exp(iw_n t)]\frac{\pi}{p} \qquad (2\text{-}14)$$

Since the term w_n is a function of the discrete variable n, let us define the first difference of w_n as

$$\Delta w = w_{n+1} - w_n = \frac{\pi(n+1)}{p} - \frac{\pi n}{p} = \frac{\pi}{p} \qquad (2\text{-}15)$$

By making that substitution in Eq. (2-14) and changing the limits of integration to the particular case of $d = -p$, we can now write

$$f_p(t) = \sum_{n=-\infty}^{\infty} F(w_n)\, \Delta w$$

where

$$F(w_n) = \frac{1}{2\pi}\exp(iw_n t)\int_{-p}^{p} f_p(t) \exp(-iw_n t)\, dt \qquad (2\text{-}16)$$

From ordinary calculus we know that under very general conditions the limit of such a sum as Δw goes to zero is an integral, namely

$$\lim_{\Delta w \to 0} f(t) = \lim_{\Delta w \to 0} \sum_{n=-\infty}^{\infty} F(w_n)\, \Delta w = \int_{-\infty}^{\infty} F(w)\, dw \qquad (2\text{-}17)$$

But Δw approaches zero as p approaches infinity. Therefore it seems reasonable to believe that the nonperiodic limit of $f(t)$ as p goes to infinity can be written as

$$f(t) = \int_{-\infty}^{\infty} \left[\frac{1}{2\pi} e^{iwt} \int_{-\infty}^{\infty} f(t) e^{-iwt}\, dt\right] dw \qquad (2\text{-}18)$$

This is the so-called *Fourier integral.* It is a valid representation of the nonperiodic function $f(t)$ so long as $f(t)$ has at most a finite number of points of discontinuity in every finite interval and the improper integral

$$\int_{-\infty}^{\infty} |f(t)|\, dt \qquad (2\text{-}19)$$

exists.

It is common practice to break the Fourier integral into two separate pieces. For example,

$$f(t) = \mathscr{F}^{-1}(F(w)) = \frac{1}{2\pi}\int_{-\infty}^{\infty} F(w) e^{iwt}\, dw \qquad (2.20)$$

and

$$F(w) = \mathscr{F}(f(t)) = \int_{-\infty}^{\infty} f(t)e^{-iwt}\,dt \tag{2-21}$$

When written in this form, $F(w)$ is referred to as the *Fourier transform* of $f(t)$ and $f(t)$ is the *inverse transform* of $F(w)$. Together they constitute a *Fourier transform pair.*

Note the characteristics of this transform pair. The function $F(w)$ represents the same phenomenon as the function $f(t)$. If $F(w)$ is known, we can always reconstruct $f(t)$ by means of Eq. (2-20). Conversely, if $f(t)$ is known, we can reconstruct $F(w)$ by means of Eq. (2-21). For that reason, which function is called the transform and which the inverse is somewhat incidental. In effect we have two representations of the same process, one in the time domain and one in the frequency domain. Our choice of which representation to use is largely one of a matter of convenience suggested by the problem context encountered. As we will soon see, transforms are often more easily manipulated than the parent functions in the context of probability modeling. Extensive tables of such transform pairs are widely available to ease the computational burden.

As an example of a Fourier transform of a function, consider the well-known negative exponential probability density function

$$f(x) = ae^{-ax}, \qquad x \geq 0$$

The Fourier transform of this function is obtained by evaluating the integral of Eq. (2-21) as follows:

$$\begin{aligned}\mathscr{F}(f(x)) &= \int_0^{\infty} ae^{-ax}e^{-iwx}\,dx = \int_0^{\infty} ae^{-(a+iw)x}\,dx \\ &= -\frac{a}{a+iw}e^{-(a+iw)x}\Big|_{x=0}^{\infty} = \frac{a}{a+iw}\end{aligned}$$

As we shall see in the next chapter, this transform can be used to recover the moments of the distribution and to develop distributions for combinations of random variables.

The Laplace Transform

The Fourier transform is a powerful theoretical tool for probability modeling because of the very general conditions under which existence is assured. On those occasions in which we are working with probability functions, the improper integral of Eq. (2-19) always exists. The integral of the density is unity. However, there are functions of interest to use which are not easily represented by the Fourier transform. For example, the simple unit step function violates the condition of convergence of the improper integral. For that reason, and also because other transforms may be easier to handle in a probability context, let us now introduce the Laplace transform and note its relationship to the Fourier transform.

Suppose that the functions of interest are all zero to the left of $t = 0$. If we attempt to obtain the Fourier transform of these functions, for some of which the improper integral of Eq. (2-19) might not exist, we find it convenient to introduce a convergence factor. To make such a technique explicit, let us introduce a surrogate function

$$g(t) = \begin{cases} 0, & t < 0 \\ e^{-at}f(t), & t \geq 0 \end{cases} \tag{2-22}$$

where $f(t)$ is the function we are actually interested in manipulating. The term e^{-at} is an artifice used to guarantee the convergence of

$$\int_0^\infty |g(t)|\ dt = \int_0^\infty |e^{-at}f(t)|\ dt \tag{2-23}$$

All that is required is that $f(t)$ be of exponential order. That is to say that the improper integral exists for some value of a or equivalently that $\lim_{t\to\infty} e^{-at}f(t) = 0$.

By applying the Fourier transform to $g(t)$ we are led to

$$\begin{aligned} \mathscr{F}(g(t)) = G(w) &= \int_0^\infty g(t)e^{-iwt}\ dt = \int_0^\infty e^{-at}f(t)e^{-iwt}\ dt \\ &= \int_0^\infty f(t)e^{-(a+iw)t}\ dt \end{aligned} \tag{2-24}$$

Furthermore, the inverse of this transform can be used to recover $f(t)$ by multiplying by e^{at}, namely

$$\begin{aligned} f(t) = e^{at}g(t) &= \frac{e^{at}}{2\pi}\int_{-\infty}^{\infty} G(w)e^{iwt}\ dw \\ &= \frac{1}{2\pi}\int_{-\infty}^{\infty} G(w)e^{(a+iw)t}\ dw \end{aligned} \tag{2-25}$$

Since both the transform and its inverse now contain the common binomial term $(a + iw)$, it is convenient to introduce a new complex variable s such that

$$s = a + iw, \qquad ds = i\ dw \tag{2-26}$$

Changing the variables of Eqs. (2-24) and (2-25), we obtain the pair

$$\mathcal{L}(f(t)) = F(s) = \int_0^\infty f(t)e^{-st}\,dt \tag{2-27}$$

and

$$F^{-1}(s) = f(t) = \frac{1}{2\pi i} \lim_{b\to\infty} \int_{a-ib}^{a+ib} F(s)e^{st}\,ds \tag{2-28}$$

This is the Laplace transform pair. Equation (2-27) is sometimes referred to as the unilateral Laplace transform since it is valid only for functions which are nonzero to the right of the origin. The bilateral Laplace transform is used for functions defined over the entire real line. In that case the lower limit of integration is minus infinity. In this book we will limit our attention to the unilateral transform. The new transform variable s is a complex variable. The inverse transform involves a contour integral, the evaluation of which requires complex variable theory. In our application we will avoid direct computation of the integral of Eq. (2-28) and rely instead on tables of transform pairs and a few basic theorems to perform any inversions required. In most of our calculations the complex variable s will be treated as a dummy variable. The functions of concern to us will be assumed to have the property of uniform convergence over the set of s-values we may need to employ. When required to evaluate particular transforms, we will normally restrict our attention to real values of s in the interval [0, 1]. Functions for which the transform exists are said to be of exponential order; that is, they do not grow more rapidly than ce^{at}, where c is a constant and the real portion of s exceeds a.

To illustrate the calculation of a Laplace transform, consider the manufacture of pistons having diameters which are uniformly distributed in the interval of 2.000 to 2.010 inches. The probability density function for piston diameters x is

$$f(x) = \begin{cases} 100, & 2.000 \le x \le 2.010 \\ 0, & \text{elsewhere} \end{cases}$$

Applying Eq. (2-27), one obtains the Laplace transform of this distribution as

$$\mathcal{L}(f(x)) = \int_0^\infty f(x)e^{-sx}\,dx = \int_{2.000}^{2.010} 100e^{-sx}\,dx$$

$$= -\frac{100}{s}e^{-sx}\Big|_{x=2.000}^{2.010} = \frac{100}{s}(e^{-2.000s} - e^{-2.010s})$$

This transform can be used to recover the moments of the distribution and to study combinations of random variables, as will be shown in Chapter 3.

The Impulse Function

The unit impulse or Dirac delta function is a useful concept in many areas of linear systems analysis. Quick disturbances or strong excitations for very brief periods may cause discontinuities in the form of spikes in system functions which require analysis. Such discontinuities or singularities can be idealized by the delta function. In addition, it is convenient to view sampling operations in which continuous waveforms are sampled at discrete points in time as a series of pulses to be described in terms of delta functions. This description is also useful for discrete probability functions.

The delta functions are not true functions in the mathematical sense. They are not defined for every value of the independent variable. They are not continuous and not differentiable. Loosely speaking, one could visualize a delta function as a function with zero width, infinite height, and an area of unity.

Symbolically, the impulse function is defined by

$$\begin{aligned} \delta(t-n) &= 0, & t &\neq n \\ \int_{-\infty}^{\infty} \delta(t-n)\,dt &= \int_{n-\varepsilon}^{n+\varepsilon} \delta(t-n)\,dt = 1, & \varepsilon &> 0 \end{aligned} \tag{2-29}$$

One could generate such a pulse mathematically in a variety of ways. For example, a triangle centered at $t = n$ with an altitude of $1/a$ and a base of $2a$ has an area of unity for all a. The limit of that function as a goes to zero would yield the properties of the delta function centered at $t = n$. In like fashion, one might view the function $\delta(t - n)$ as the limit of a normal probability distribution with a mean of n and standard deviation σ as σ approaches zero. Such functions are represented graphically as vertical arrows with the height of the arrow being the magnitude of the pulse.

The most important property of the delta function is its so-called sifting property. It can be used to select particular values of a function $f(t)$ by

$$\int_{-\infty}^{\infty} \delta(t-n) f(t)\,dt = f(n) \tag{2-30}$$

In this context the entire left-hand member is treated as a single operation. The practical importance of this operation will become apparent as the development of transfer functions for linear systems is presented.

The z-Transform

Let us now consider a sampled data system in which impulses are generated at $t = nT$ with strengths $f(nT)$. The variable T is the sampling interval, which in many applications will be unity. The output from this system is the sampled function $f_1(t)$, where

$$f_1(t) = \sum_{n=0}^{\infty} f(nT)\,\delta(t - nT) \tag{2-31}$$

This operation is illustrated in Figure 2-4.

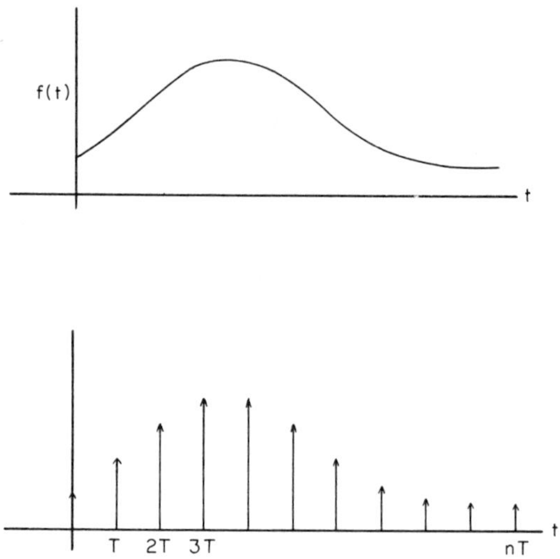

Figure 2-4 Sampled data function.

Now suppose that it is desirable to find the Laplace transform of the sampled function $f_1(t)$. By definition

$$\begin{aligned}\mathscr{L}(f_1(t)) &= \int_0^{\infty} f_1(t)e^{-st}\,dt \\ &= \int_0^{\infty} \sum_{n=0}^{\infty} f(nT)\,\delta(t - nT)e^{-st}\,dt \end{aligned} \tag{2-32}$$

However, because of the peculiar properties of the delta function, this integral reduces to the summation

$$\mathscr{L}(f_1(t)) = \sum_{n=0}^{\infty} f(nT)e^{-snT} \tag{2-33}$$

Most of our discrete functions can be viewed as having a sampling interval of $T = 1$. Furthermore, it is convenient to introduce a new variable z in the transform of discrete functions, where

$$z = e^{sT}$$

The transform of the sampled function is then given the name *z-transform*, defined as

$$Z(f(n)) = \sum_{n=0}^{\infty} f(n)z^{-n} \tag{2-34}$$

Note that this transformation, although developed in the context of sampled data systems, is a perfectly valid transform for any discrete function $f(n)$. In our applications $f(n)$ may be a probability distribution for an integer-valued random variable rather than a sampled function per se. In either case the z-transform can be viewed as the discrete version of the Laplace transform. As such it has all of the power and limitations inherent in the Laplace transform. We will employ long division, partial fraction expansion, etc. in preference to executing the complex inversion formula when inverting the transform.

To illustrate the calculation of z-transforms, consider the final inspection station at an automobile assembly plant. The quality control department has established that the number of paint defects per automobile hood can be described by a Poisson distribution with a mean of three per hood. They need to find the z-transform of this distribution in order to complete an economic model related to customer complaints. The function to be transformed is

$$f(n) = \frac{3^n e^{-3}}{n!}, \qquad n = 0, 1, 2, \dots$$

Applying Eq. (2-34), we obtain

$$Z(f(n)) = \sum_{n=0}^{\infty} \frac{3^n e^{-3}}{n!} z^{-n} = e^{-3} \sum_{n=0}^{\infty} \left(\frac{3}{z}\right)^n \Big/ n!$$

The infinite series is of the form of a power series expansion of e^{at} where $a = 3/z$. It therefore follows that

$$Z(f(n)) = e^{-3} e^{3/z} = e^{(3/z)(1-z)}$$

As with the Fourier and Laplace transforms of probability functions, this z-transform can now be used to find the moments of the distribution and to study combinations of random variables.

The Exponential and Geometric Transforms

In much of the literature relating to probability modeling slight variations of the Laplace transform and the related z-transform are employed. We will refer to these variations as the *exponential transform* and the *geometric transform*, respectively. They are also frequently referred to as moment generating functions and probability generating functions, for reasons which will become clear in later chapters.

The exponential transform looks like a Laplace transform in which the exponential term has a positive instead of a negative exponent. Furthermore, the transform is not usually restricted to functions defined only to the right of zero. We will define the exponential transform of $f(t)$ as

$$M(s) = \int_{-\infty}^{\infty} f(t)e^{st}\,dt \tag{2-35}$$

Although this transform works well for many probability distributions, there are functions for which the exponential transform does not exist. Furthermore, it is not well tabulated outside the area of probability modeling. For these reasons we will use the Laplace transform in preference to the exponential transform whenever possible.

The geometric transform plays the same role vis-à-vis the exponential transform as did the z-transform vis-à-vis the Laplace transform. The geometric transform looks like a z-transform in which the z-variable carries a positive instead of a negative exponent. Furthermore, the limits can in theory be extended from minus to plus infinity. However, the tables in the Appendices are based on transforms defined only for positive values of n. We will define the geometric transform as

$$G(z) = \sum_{n=-\infty}^{\infty} f(n)z^n \tag{2-36}$$

Consider a discrete linear control system which is being subjected to a driving function of the form

$$f(n) = (0.5)^n, \qquad n = 0, 1, 2$$

In order to develop the transfer function for this system, it is necessary to find the geometric transform of $f(n)$. From Eq. (2-36)

$$G(f(n)) = \sum_{n=0}^{\infty} (0.5)^n z^n = \sum_{n=0}^{\infty} (0.5z)^n$$

If we restrict z to being less than two, this series is of the same form as the familiar geometric series with the closed form

$$\sum_{i=0}^{\infty} (r)^i = \frac{1}{1-r}$$

It therefore follows that

$$G(f(n)) = \frac{1}{1-0.5z}$$

The Mellin Transform

One other integral transform which deserves mention is the Mellin transform. This transform is sometimes useful in studying the distribution of products and quotients of independent random variables. The Mellin transform of a function $f(x)$, where x is positive, is defined as

$$F_{\mathrm{M}}(s) = \int_0^{\infty} f(x)x^{s-1}\,dx \tag{2-37}$$

Epstein has also extended this technique to random variables which are not everywhere positive [12]. We shall have much more to say about this transform when we discuss combinations of random variables.

Table 2-1

Transform Definitions

$f(t),\quad -\infty < t < \infty$ continuous	Fourier	$F(w) = \int_{-\infty}^{\infty} f(t)e^{-iwt}\,dt$
$f(t),\quad 0 < t < \infty$ continuous	Laplace	$F(s) = \int_0^{\infty} f(t)e^{-st}\,dt$
$f(t),\quad \infty < t < \infty$ continuous	Exponential	$M(s) = \int_{-\infty}^{\infty} f(t)e^{st}\,dt$
$f(n),\quad n = 0, 1, 2, \ldots$ discrete	z	$Z(z) = \sum_{n=0}^{\infty} f(n)z^{-n}$
$f(n),\quad n = \ldots, -2, -1, 0, 1, 2, \ldots$ discrete	Geometric	$G(z) = \sum_{n=-\infty}^{\infty} f(n)z^{n}$
$f(x),\quad 0 < x < \infty$ continuous	Mellin	$F_M(s) = \int_0^{\infty} f(x)x^{s-1}\,dx$

Suppose that we have a random variable x which is uniformly distributed over the interval $[a, b]$. The Mellin transform for the rectangular distribution

$$f(x) = \begin{cases} \dfrac{1}{b-a}, & a \le x \le b \\ 0, & \text{elsewhere} \end{cases}$$

is given by

$$F_{\mathrm{M}}(s) = \int_a^b \frac{1}{b-a} x^{s-1}\,dx = \frac{x^s}{s(b-a)}\bigg|_a^b = \frac{b^s - a^s}{s(b-a)}$$

This transform can now be used with others to study the product of x and other random variables.

The definitions of the various transforms and their major properties are summarized in Table 2-1.

Fundamentals of Laplace Transforms

Many of the techniques associated with manipulating most of the commonly used integral transformations can be illustrated by examining the characteristics of one such transform. For that reason, plus the fact that it is widely available in tabulated form, we will concentrate our presentation of transforms for continuous functions on the Laplace transform.

Transform Pairs

Transforms for simple functions $f(t)$ can be developed from a straightforward application of the definition of the Laplace transform. For example, consider the unit step function at the origin

$$U(t) = \begin{cases} 0, & t < 0 \\ 1, & t \geq 0 \end{cases}$$

The Laplace transform of $U(t)$ is given by

$$\mathscr{L}(U(t)) = \int_0^\infty 1 \cdot e^{-st}\,dt = -\frac{1}{s}e^{-st}\bigg|_0^\infty = \frac{1}{s} \tag{2-38}$$

This now gives us a pair of functions $U(t)$ and $\mathscr{L}(U(t))$ which we can tabulate for future reference. If, then, during the course of calculation we find it necessary to invert a transform $F(s) = 1/s$, it will be unnecessary to employ the complex inversion formula. Rather, we will merely consult the tables and find that $F^{-1}(1/s) = U(t)$ as demonstrated above.

In similar fashion we can develop the transform of

$$f(t) = t, \qquad t > 0$$

by using integration by parts. That is, if

$$\mathscr{L}(f(t)) = \int_0^\infty te^{-st}\,dt$$

let

$$u = t, \qquad du = dt, \qquad dv = e^{-st}\,dt, \qquad v = \frac{-e^{-st}}{s}$$

Then

$$\mathscr{L}(t) = uv - \int v\,du = -\frac{te^{-st}}{s}\bigg|_0^\infty + \frac{1}{s}\int_0^\infty e^{-st}\,dt = \frac{1}{s^2} \tag{2-39}$$

This transform pair can now also be entered in our table of transform pairs for future reference. The development of transforms for other elementary functions will be left as exercises for the reader.

To illustrate how one might exploit the resulting table of transform pairs, consider the following examples. Let

$$f(t) = t^2 + 4t - 3\sin 5t$$

Since the Laplace transform is a linear operator, we can immediately break the problem into three subproblems,

$$\mathscr{L}(f(t)) = \mathscr{L}(t^2) + 4\mathscr{L}(t) - 3\mathscr{L}(\sin 5t)$$

From the table of transform pairs in Appendix B we find

$$\mathscr{L}(t^2) = \frac{2}{s^3}, \qquad \mathscr{L}(t) = \frac{1}{s^2}, \qquad \mathscr{L}(\sin 5t) = \frac{5}{s^2 + 25}$$

It therefore follows that

$$\mathscr{L}(f(t)) = \frac{2}{s^3} + \frac{4}{s^2} - \frac{15}{s^2 + 25}$$

Note that we did not have to evaluate a single integral to obtain this transform.

Looking at the converse problem, suppose that we have found the transform of $f(t)$ to be

$$\mathscr{L}(f(t)) = \frac{12}{s^4} + \frac{5}{s^2} + \frac{6s}{s^2 + 16}$$

Again using the linear property of Laplace transforms, let us rewrite the function as

$$\mathscr{L}^{-1}(f(t)) = 2\mathscr{L}^{-1}\left(\frac{6}{s^4}\right) + 5\mathscr{L}^{-1}\left(\frac{1}{s^2}\right) + 6\mathscr{L}^{-1}\left(\frac{s}{s^2 + 4^2}\right)$$

From the table of transform pairs we find that

$$\mathscr{L}^{-1}\left(\frac{n!}{s^{n+1}}\right) = t^n, \qquad \mathscr{L}^{-1}\left(\frac{1}{s^2}\right) = t, \qquad \mathscr{L}^{-1}\left(\frac{s}{s^2 + w^2}\right) = \cos(wt)$$

It therefore follows that

$$f(t) = 2t^3 + 5t + 6 \cos 4t$$

Again the function is established without resorting to the complex inversion formula.

Partial Fraction Expansion

Since our available tables consist of relatively simple functions, we now need some method for breaking a transform into simple parts so that we can use the linearity property and brief tables to find the inverse. One useful technique for doing this is partial fraction expansion.

In many applications a transform solution will take the form of a quotient of polynomials

$$F(s) = \frac{A(s)}{B(s)} \tag{2-40}$$

Here $A(s)$ and $B(s)$ are polynomials in s with the degree of the numerator assumed to be less than the degree of the denominator. If the ratio is not a proper rational fraction, it can be replaced by two terms, one of which is such a fraction, by division.

When the denominator can be factored into a product of linear terms, we know from algebra that the fraction can be expressed as a sum of separate terms each having as its denominator a factor of the original denominator. The values of the numerators are unknowns, yet to be determined. For example, suppose that we must invert the following transform:

$$\mathscr{L}(f(t)) = \frac{s+1}{s^3 - 4s} = \frac{s+1}{s(s+2)(s-2)}$$

By partial fractions,

$$\frac{s+1}{s(s+2)(s-2)} = \frac{c_1}{s} + \frac{c_2}{s+2} + \frac{c_3}{s-2}$$

where c_1, c_2, and c_3 must be determined.

When the denominator can be factored into n distinct roots, the transform is said to have n simple or *first-order poles*. Roots of the numerator are called *zeros*. In the example above, the first-order poles are $s = 0$, $s = -2$, $s = 2$. The zero is $s = -1$. These are simply the values of s that make the denominator or numerator zero. When the denominator has only first-order poles a_k, the coefficients c_k can be calculated from a simple limit argument.

Note that multiplying by $(s - a_k)$ will bring c_k standing alone on the right, namely

$$(s - a_k)F(s) = \frac{(s - a_k)c_1}{s - a_1} + \cdots + c_k + \frac{(s - a_k)c_{k+1}}{s - a_{k+1}} + \cdots$$

If we now take the limit of both sides as s approaches a_k, all terms except c_k on the right will go to zero. Therefore

$$c_k = \lim_{s \to a_k} (s - a_k)F(s) \tag{2-41}$$

for all first-order poles.

Applying this technique to our example, we obtain

$$c_1 = \lim_{s \to 0} s \frac{s + 1}{s(s + 2)(s - 2)} = -\frac{1}{4}$$

$$c_2 = \lim_{s \to -2} (s + 2) \frac{s + 1}{s(s + 2)(s - 2)} = -\frac{1}{8}$$

$$c_3 = \lim_{s \to 2} (s - 2) \frac{s + 1}{s(s + 2)(s - 2)} = \frac{3}{8}$$

The transform in expanded form can now be written as

$$\mathscr{L}(f(t)) = \frac{-1/4}{s} + \frac{-1/8}{s + 2} + \frac{3/8}{s - 2}$$

It is now a simple matter to consult the tables of Appendix B to find that the inverse is

$$f(t) = -\tfrac{1}{4} - \tfrac{1}{8}e^{-2t} + \tfrac{3}{8}e^{2t}$$

When a factor in the denominator is repeated k times, the transform is said to have a kth-order pole. Each such pole will contribute k terms to the partial fraction expansion. For example, suppose that the kth-order pole is at a_1. The partial fraction expansion would then contain k fractions associated with that pole plus as many more fractions as there are first-order poles, namely

$$F(s) = \frac{A(s)}{(s - a_1)^k (s - a_2)(s - a_3) \cdots}$$

$$= \frac{c_{11}}{(s - a_1)^k} + \frac{c_{12}}{(s - a_1)^{k-1}} + \cdots \frac{c_{1k}}{s - a_1} + \frac{c_2}{s - a_2} + \frac{c_3}{s - a_3} + \cdots \tag{2-42}$$

Here the double subscript on c_{ij} is used to denote the jth term associated with the ith higher order pole.

The technique for determining the numerators in the expansion must now be modified somewhat. If we multiply the transform by $(s - a_1)^k$, all terms other than c_{11} will have some multiple of $(s - a_1)$ as a coefficient, which in the limit will go to zero. However, if we multiply by $(s - a_1)^{k-1}$, the first term will contain $(s - a_1)$ in the denominator, which will cause difficulties as s approaches a_1. This difficulty is circumvented by first multiplying $F(s)$ by $(s - a_1)^k$ and then taking an appropriate number of derivatives with respect to the s before evaluating the limit. In general the jth coefficient for terms resulting from a kth-order repeated root a_i can be obtained from

$$c_{ij} = \frac{1}{(j-1)!} \lim_{s \to a_i} \frac{d^{j-1}}{ds^{j-1}} (s - a_i)^k F(s) \tag{2-43}$$

where $0 < j \leq k$, with k the order of the pole under investigation.

To illustrate the technique, suppose that the transform in our previous example had been

$$F(s) = \frac{s+1}{s(s+2)(s-2)^2}$$

We now have a second-order pole at $s = 2$ and first-order poles at $s = 0$ and $s = -2$. In expanded form

$$F(s) = \frac{c_1}{s} + \frac{c_2}{s+2} + \frac{c_{31}}{(s-2)^2} + \frac{c_{32}}{s-2}$$

The coefficients for the first-order terms are determined as before from Eq. (2-41):

$$c_1 = \lim_{s \to 0} sF(s) = \tfrac{1}{8}, \qquad c_2 = \lim_{s \to -2} (s+2)F(s) = \tfrac{1}{32}$$

The coefficients for the second-order term are

$$c_{31} = \lim_{s \to 2} (s-2)^2 F(s) = \frac{3}{8}, \qquad c_{32} = \lim_{s \to 2} \frac{d}{ds} (s-2)^2 F(s) = -\frac{5}{32}$$

The transform is then

$$F(s) = \frac{1/8}{s} + \frac{1/32}{s+2} + \frac{3/8}{(s-2)^2} - \frac{5/32}{(s-2)}$$

which can now be inverted by table lookup.

Method of Residues

A method for finding inverse transforms which closely resembles partial fraction expansion is the method of residues. The major differences are that an nth-order pole has a single residue associated with it and that the residue theorem yields the inverse directly without a table lookup.

The residue is a number arising from a series expansion of a function of a complex variable.[2] When we want to invert a Laplace transform that can be expressed as a rational fraction $F(s)$ the method of residues states that

$$\mathscr{L}^{-1}(F(s)) = \sum_{\text{all poles}} [\text{residues of } F(s)e^{st}] \tag{2-44}$$

where the residue at an nth-order pole $s = a$ is given by

$$R_a = \frac{1}{(n-1)!} \frac{d^{n-1}}{ds^{n-1}} [(s-a)^n F(s) e^{st}] \bigg|_{s=a} \tag{2-45}$$

To illustrate the technique, let us use the method of residues to find the inverse of

$$F(s) = \frac{1}{(s-a)^3}$$

This transform has a single third-order pole at $s = a$. Therefore application of Eqs. (2-44) and (2-45) involves the second derivative of $F(s)e^{st}$, namely

$$R_a = \frac{1}{(3-1)!} \frac{d^2}{ds^2} \left[(s-a)^3 \frac{1}{(s-a)^3} e^{st} \right] \bigg|_{s=a} = \frac{t^2 e^{st}}{2} \bigg|_{s=a}$$

from which

$$f(t) = \tfrac{1}{2} t^2 e^{at}$$

The fact that this is the function with the transform $1/(s-a)^3$ is easily verified by checking the tables. Note that the method of partial fractions offers no assistance in this instance because the transform cannot be further reduced.

Even though the method of residues yields the inverse directly, it offers little savings in effort from partial fraction expansion when both are applicable. The proper choice of technique is largely one of personal preference.

[2] For a discussion of residues see Churchill [4, pp. 206–213].

s-Plane Analysis

Many problems can be solved without the sometimes tedious task of inverting the transform. This is especially true in the design of physical systems where questions of stability are of utmost importance. The method for accomplishing this is s-plane analysis.

The s-plane is formed by axes identified as the imaginary and the real axes. Zeros and poles can be plotted as vectors in the s-plane. In our previous examples the zeros and poles had no imaginary components, but in general one can expect the roots to be complex numbers of the form

$$s = a \pm iw \tag{2-46}$$

Furthermore, we know from complex variable theory that such roots always appear in complex conjugate pairs. What we now need are simple procedures for manipulating complex numbers.

When working with complex numbers it is often easiest to express them as exponential terms. From DeMoivre's theorem it is well known that

$$e^{ix} = \cos x + i \sin x \tag{2-47}$$

The exponential quantity can be interpreted as a unit vector making an angle x with the real axis as in Figure 2-5a. Thus the complex number

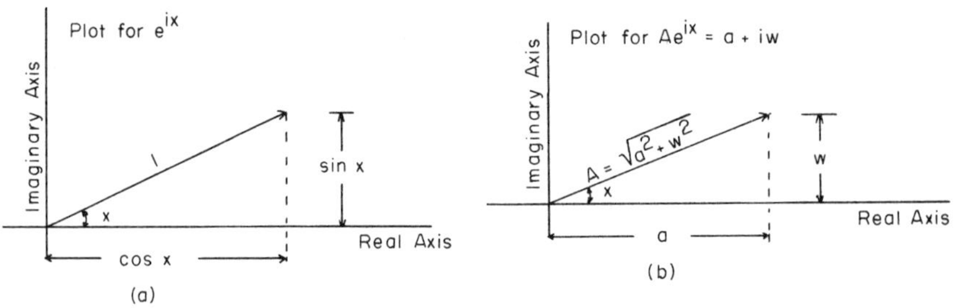

Figure 2-5 Plotting complex numbers in the s-plane.

plotted as a vector in Figure 2-5b can be expressed in exponential form as

$$a + iw = Ae^{ix} \tag{2-48}$$

where

$$A = (a^2 + w^2)^{1/2} \qquad \text{and} \qquad x = \tan^{-1} \frac{w}{a}$$

The variable A is referred to as the amplitude and x represents the phase.

Consider now the significance one can attach to the poles of a transform when plotted as vectors in the s-plane. Poles with no imaginary component can be plotted as vectors along the real axis. If the pole is to the right of the imaginary axis, the inverse transform associated with that pole is

$$\mathscr{L}^{-1}\left(\frac{A}{s-a}\right) = Ae^{at} \tag{2-49a}$$

The time function for a positive real pole is an increasing exponential. If the pole lies to the left of the imaginary axis, the inverse transform is

$$\mathscr{L}^{-1}\left(\frac{A}{s+a}\right) = Ae^{-at} \tag{2-49b}$$

which is a decaying exponential.

Poles with no real component appear as vectors on the imaginary axis. The inverse transform for such terms would lead to constant-amplitude sinusoidal oscillation. Poles with both real and imaginary components in the left half of the s-plane lead to damped sinusoids, while those in the right half result in exponentially increasing terms which are unstable. Higher order poles lead to additional terms involving powers of t but with the same general stability properties as those noted for first-order poles. The only exception is for higher order poles lying on the imaginary axis. They lead to unstable sinusoids.

It is thus possible to establish the general character of a system's response from the s-plane plot without carrying out the transform inversion.

Transforms of Derivatives

The reason for transforming functions from the time domain to the frequency domain is to obtain computational advantages. Let us now turn our attention to a few of the fundamental properties of Laplace transforms which assure their usefulness in computational problems.

To begin with, consider the transform of the derivative of a continuous function $f(t)$. From the definition and applying integration by parts we have

$$\mathscr{L}(f'(t)) = \int_0^\infty e^{-st}f'(t)\,dt = e^{-st}f(t)\Big|_0^\infty + s\int_0^\infty e^{-st}f(t)\,dt$$

The first term on the right reduces to the value of the function $f(t)$ at $t = 0$. The second involves an integral which is easily recognized as the Laplace transform of $f(t)$. It then follows that

$$\mathscr{L}(f'(t)) = s\mathscr{L}(f(t)) - f(0) \tag{2-50}$$

The above property is sometimes referred to as the fundamental operational property of the Laplace transformation. By using this property we can replace the operation of differentiation by a simple algebraic operation on the transform. Note that this property holds only if $f(t)$ is continuous and of exponential order.

One can continue a similar development for higher order derivatives. For example,

$$\mathscr{L}(f''(t)) = e^{-st}f'(t)\Big|_0^\infty + s\int_0^\infty e^{-st}f'(t)\,dt = -f'(0) + s\mathscr{L}(f'(t))$$

from which

$$\mathscr{L}(f''(t)) = s^2\mathscr{L}(f(t)) - sf(0) - f'(0) \tag{2-51}$$

In similar fashion one can establish an equation for the nth derivative as

$$\mathscr{L}(f^{(n)}(t)) = s^n\mathscr{L}(f(t)) - s^{n-1}f(0) - \cdots - sf^{(n-2)}(0) - f^{(n-1)}(0) \tag{2-52}$$

The simplification obtained from this property is most appreciated in the solution of high-order differential equations. In addition, Eq. (2-52) can be used to develop more transform pairs for our table. For example, consider the function $f(t) = t^r$ for any positive integer r, which is continuous and of exponential order. The conditions at the boundary are simply

$$f^{(n)}(0) = \begin{cases} 0 & \text{for all } n \neq r \\ r! & \text{for } n = r \end{cases}$$

Applying Eq. (2-52) for $n = r + 1$, we have

$$\mathscr{L}(f^{(r+1)}(t)) = 0 = s^{r+1}\mathscr{L}(t^r) - r!$$

from which

$$\mathscr{L}(t^r) = \frac{r!}{s^{r+1}} \tag{2-53}$$

In the event that the exponent on t is not restricted to being an integer, the formula can be generalized to

$$\mathscr{L}(t^k) = \frac{\Gamma(k+1)}{s^{k+1}}, \qquad k > -1 \tag{2-54}$$

where the gamma function is defined as

$$\Gamma(k+1) = \int_0^\infty x^k e^{-x}\,dx \tag{2-55}$$

For details of this development and many others see Churchill [4, p. 11].

Initial and Final Value Theorems

Limited information concerning system behavior can be obtained directly from the transform by use of the initial and final value theorems, which are also easily proven from the properties of the transform of a derivative.

The initial value theorem states that the behavior of a time function $f(t)$ in the neighborhood of $t = 0$ corresponds to the behavior of $s\mathscr{L}(f(t))$ in the neighborhood of $s = \infty$. Proof of this assertion begins with the equation for the transform of the derivative $f'(t)$. From our previous work on the transform of a derivative we have

$$\mathscr{L}(f'(t)) = \int_0^\infty e^{-st}f'(t)\,dt = sF(s) - f(0)$$

The limit of both the right- and left-hand sides as s goes to infinity yields

$$0 = \lim_{s\to\infty} [sF(s) - f(0)]$$

from which

$$\lim_{s\to\infty} sF(s) = \lim_{t\to 0} f(t) \tag{2-56}$$

This is the initial value theorem.

The final value theorem states that

$$\lim_{s\to 0} sF(s) = \lim_{t\to\infty} f(t)$$

The proof for the theorem also begins with the expression for the transform of a derivative. Details of that proof are left as an exercise for the reader.

When applicable, these theorems permit the analyst to establish the value of the time function at the origin and the limiting value of the function directly from the transform without going through the inversion process. The restrictions on using the final value theorem are that $f(t)$ and $f'(t)$ must be Laplace-transformable as t goes to infinity and that the poles of $F(s)$, excluding the origin, must lie in the left half of the s-plane. Otherwise the time function is an oscillating function which has no limit. It is always wise to check the location of the poles before applying the final value theorem to avoid the possibility of spurious results.

To illustrate the technique, suppose that the transform for a system response function has been given as

$$F(s) = \frac{m(32 + 3s)}{s(ms + 5)} \tag{2-57}$$

This function has first-order poles at $s = 0$ and $s = -5/m$, both of which are admissible for using the theorems. From the initial value theorem

$$f(0) = \lim_{s\to\infty} sF(s) = \lim_{s\to\infty} \frac{m(32 + 3s)}{ms + 5} = 3$$

and from the final value theorem

$$\lim_{t\to\infty} f(t) = \lim_{s\to 0} \frac{m(32 + 3s)}{ms + 5} = \frac{32m}{5}$$

If we only need the limiting behavior of this system, it is unnecessary to go through the inversion process.

Integration by Transforms

The Laplace transform can be used to simplify the operation of integration of well-behaved functions. Integration in the time domain reduces to division by s in the frequency domain. The relationship, which follows directly from the definition, is

$$\mathscr{L}\left(\int_0^t f(t)\,dt\right) = \frac{1}{s}\mathscr{L}(f(t)) \tag{2-58}$$

This technique is particularly valuable in probability modeling, where we often need to work with cumulative probability functions. By using the integration theorem the transform of the cumulative distribution is readily available without first executing a sometimes difficult integration in time domain and then transforming. Much more will be said about this technique in later chapters.

Shifting Theorems

Two shifting properties of transforms are important to us. In the first case, a shift of a units on the time axis can be accomplished in the transform domain by multiplying the transform of a function with its origin at $t = 0$ by e^{-as}. That is,

$$\mathscr{L}(f(t - a)U(t - a)) = e^{-as}F(s) \tag{2-59}$$

where $F(s)$ is the transform of the unshifted function. The unit step function is included to ensure that the nonnegative restriction on the Laplace transform is not violated. This theorem answers the question of how the transform changes when the t argument of the function is replaced by $t - a$.

A corresponding result applies when one seeks changes in the time function caused by replacing the transform argument by $s - a$. This is referred to as complex shifting. The theorem states that

$$\mathscr{L}[e^{at}f(t)] = F(s - a) \tag{2-60}$$

where $F(s)$ is the transform of $f(t)$. If one wants to multiply a function with a known transform by e^{at}, it is then a simple matter to obtain the new transform. For example,

$$\mathscr{L}(te^{4t}) = \mathscr{L}(t)\Big|_{s=s-4} = \frac{1}{s^2}\Big|_{s-4} = \frac{1}{(s-4)^2}$$

This is a very handy device for expanding simple tables to find transforms of more complex functions.

Multiplication and Division

In previous discussions it was noted that integration and differentiation of functions could be accomplished in the transform domain by multiplication and division of the transform by s. Multiplication and division of $f(t)$ can be accomplished in similar fashion by integration and differentiation of $\mathscr{L}(f(t))$. Let $\mathscr{L}(f(t)) = F(s)$; then

$$\mathscr{L}(tf(t)) = -\frac{d}{ds}F(s) \tag{2-61}$$

and

$$\mathscr{L}\left(\frac{1}{t}f(t)\right) = \int_s^{\infty} F(s)\,ds \tag{2-62}$$

Proof of these relationships follows in straightforward fashion by performing the indicated operations on $F(s)$.

As before one can use these theorems to evaluate transforms of functions containing components $f(t)$ whose transforms are already known.

Solving Differential and Integrodifferential Equations

Most of the preceding discussion of Laplace transforms has been limited to the mechanics of obtaining the transforms of given functions and inverting specified transforms to obtain the corresponding functions of time. We are now in a position to use these skills in the solution of more meaningful problems, such as those posed by differential equations.

Consider, for example, the following linear differential equation for a simple RL circuit:

$$L\frac{dI(t)}{dt} + RI(t) = E(t) \tag{2-63}$$

Suppose the system, initially at rest, is subjected to the constant driving force $E(t) = k$. The problem is to determine the flow of current $I(t)$ as a function of time.

The first step is to form the *subsidiary equation*. The subsidiary equation is an equation in transforms which arises from the given differential equation. By applying the Laplace transform, ordinary differential equations are reduced to algebraic equations of the transform. By transforming both sides of Eq. (2-63), we have

$$\mathscr{L}\left[L\frac{dI(t)}{dt} + RI(t)\right] = \mathscr{L}[k]$$

or

$$L[s\mathscr{L}(I(t)) - I(0)] + R\mathscr{L}(I(t)) = \frac{k}{s}$$

Solving for the unknown transform,

$$\mathscr{L}(I(t)) = \frac{k}{s(Ls + R)}$$

By partial fractions this can be rewritten as

$$\mathscr{L}(I(t)) = \frac{c_1}{s} + \frac{c_2}{s + R/L} = \left[\frac{1}{s} - \frac{1}{s + R/L}\right]\frac{k}{R}$$

From the table of transform pairs it follows that

$$I(t) = \frac{k}{R}\left(1 - \exp\left(-\frac{Rt}{L}\right)\right) \tag{2-64}$$

Laplace transforms are also quite useful in solving certain classes of integral equations and integrodifferential equations. (An integral equation is one in which the unknown function occurs inside an integral.) For example, suppose that we expand the circuit equation used above to include a term for voltage drop across a condenser. The equation now becomes an integrodifferential equation in which the unknown current flow appears in both derivative and integral terms, namely

$$L\frac{dI(t)}{dt} + RI(t) + \frac{1}{C}\int_0^t I(t)\,dt = k \tag{2-65}$$

The transform of the integral terms is, from Eq. (2-58),

$$\mathscr{L}\left(\frac{1}{C}\int_0^t I(t)\,dt\right) = \frac{1}{Cs}\mathscr{L}(I(t))$$

The subsidiary equation now becomes

$$L[s\mathscr{L}(I(t)) - I(0)] + R\mathscr{L}(I(t)) + \frac{1}{Cs}\mathscr{L}(I(t)) = \frac{k}{s}$$

Solving for the transform of the unknown function,

$$\mathscr{L}(I(t)) = \frac{kC}{LCs^2 + RCs + 1}$$

which can be inverted by the usual methods to recover the function $I(t)$.

Another major class of integral equations is that of the convolution type. The general form for such equations is

$$f(t) = g(t) + \int_0^t h(t - y)f(y)\,dy \tag{2-66}$$

where the g and h functions are known and $f(t)$ is to be found. Transforming both sides yields

$$\mathscr{L}(f(t)) = \mathscr{L}(g(t)) + \mathscr{L}(h(t))\mathscr{L}(f(t))$$

from which

$$\mathscr{L}(f(t)) = \frac{\mathscr{L}(g(t))}{1 - \mathscr{L}(h(t))}$$

We will make extensive use of the convolution integral and its accompanying transform properties in later discussions involving sums of random variables.

A Caveat We have presented many diverse facts about Laplace transforms without great care in following their mathematical development and noting the subtleties implied by their use. We assume that the integral converges for all s-values of interest. The rules presented work well for most well-behaved functions. However, when problems are encountered which will not yield to the tools discussed, the reader is urged to consult a higher level mathematics text containing the necessary rigor. We are using a pragmatic approach. We employ transform techniques as though they are always valid and fall back on more rigorous analysis only when answers are impossible to obtain by these techniques or when they violate our intuition.

Fundamentals of Discrete Transforms

The z- and geometric transforms can be used for discrete time functions much as we use Laplace and exponential transforms for continuous time functions. Indeed, one might describe their use as techniques for finite operational calculus. This means that they have the fundamental property of transforming linear difference equations into algebraic equations in the transform. In addition, they have properties involving summation of discrete functions, shifting theorems, initial and final value theorems, etc. In other words they provide the same operational capabilities for functions with integer arguments that Laplace transforms provide for continuous functions. Because their operational properties are so similar to the Laplace transform, the following discussion will be brief.

Transform Pairs

As with all transforms, finding the z- or geometric transform of a function and its inverse is most easily accomplished by using certain fundamental properties of the transform and a table of transform pairs.

Whether or not a particular function has a discrete transform depends upon the convergence of the power series defining the transform. A common test for convergence is the ratio test, which gives a sufficient condition for convergence. The ratio test is based on the following theorem.

Theorem Given the series $\sum_{i=0}^{\infty} u_i$, find the limit

$$L = \lim_{n\to\infty} \frac{u_{n+1}}{u_n}$$

If this limit is less than unity, the series is absolutely convergent. If this limit is greater than unity, the series is divergent. If the limit equals unity, no conclusions can be drawn.

As an example of the usefulness of this test, suppose that we are asked to find the geometric transform of

$$f(n) = n!$$

By definition

$$G(f(n)) = \sum_{n=0}^{\infty} f(n)z^n = \sum_{n=0}^{\infty} n!\, z^n$$

However, from the ratio test

$$L = \lim_{n\to\infty} \frac{u_{n+1}}{u_n} = \lim_{n\to\infty} \frac{(n+1)!\, z^{n+1}}{n!\, z^n} = \lim_{n\to\infty} (n+1)z$$

But this limit exists only for $z = 0$. It therefore follows that the function $f(n) = n!$ does not possess a geometric transform if we permit n to go to infinity.

It should be noted that if n is limited to some finite number in the above transform, the geometric transform is well defined. The finite sum

$$G(f(n)) = \sum_{n=0}^{m} f(n)z^n$$

is well defined. It is only when an infinite series is involved that questions of convergence become critical. For most common functions we will encounter, both finite and infinite, convergence will be assumed without test. Only when difficulties are encountered will we resort to an explicit test of convergence.

Tables of transform pairs are constructed by repeated application of the definition. For example, consider the z-transform of

$$f(n) = \begin{cases} c, & n \geq 0 \\ 0, & n < 0 \end{cases}$$

By definition

$$Z(f(n)) = \sum_{n=0}^{\infty} cz^{-n} = c(1 + z^{-1} + z^{-2} + \cdots)$$

The infinite sum in parentheses is a geometric series with an argument $r = z^{-1}$. As long as z is greater than unity, it follows that

$$\sum_{i=0}^{\infty} (z^{-1})^i = \frac{1}{1 - z^{-1}} = \frac{z}{z - 1}$$

from which

$$Z(c) = \frac{cz}{z - 1} \tag{2-67}$$

We now have established one entry for our table of transform pairs.

Transforms for other common functions are proven in a straightforward fashion. A brief table of transform pairs is presented in Appendix C for the convenience of the reader.

Operating Rules

Discrete transforms gain their power from certain operating characteristics which permit simplification of linear difference equations and certain problems in sum calculus. The computational advantage is of similar magnitude to that demonstrated for Laplace transforms.

For example, consider a simple shifting operation using geometric transforms. By definition

$$G(f(n+1)) = \sum_{n=0}^{\infty} f(n+1)z^n = z^{-1}\sum_{n=0}^{\infty} f(n+1)z^{n+1}$$

$$= z^{-1}\sum_{k=1}^{\infty} f(k)z^k = z^{-1}[G(f(n)) - f(0)] \qquad (2\text{-}68)$$

From this shifting property, it follows that the geometric transform of the first forward difference of $f(n)$ is

$$\begin{aligned} G(\Delta f(n)) &= G(f(n+1) - f(n)) \\ &= z^{-1}[G(f(n)) - f(0)] - G(f(n)) \\ &= (z^{-1} - 1)G(f(n)) - z^{-1}f(0) \end{aligned} \qquad (2\text{-}69)$$

A similar result is easily shown for the z-transform. That is,

$$Z(\Delta f(n)) = (z - 1)Z(f(n)) - zf(0) \qquad (2\text{-}70)$$

follows directly from the definitions. The resemblance to the Laplace transform for the derivative of continuous functions is striking.

There is an initial value theorem and a final value theorem for each of the discrete transforms for use whenever only the limits of a function are needed. The initial value theorems are

$$f(0) = \lim_{z\to 0} G(f(n)) \qquad (2\text{-}71)$$

for geometric transforms and

$$f(0) = \lim_{z\to\infty} Z(f(n)) \qquad (2\text{-}72)$$

for z-transforms. Both follow directly from the definitions. The final value theorems are

$$\lim_{n\to\infty} f(n) = \lim_{z\to 1} (1 - z)G(f(n)) \qquad (2\text{-}73)$$

and

$$\lim_{n\to\infty} f(n) = \lim_{z\to 1} \left(\frac{z-1}{z}\right) Z(f(n)) \qquad (2\text{-}74)$$

For these theorems to have meaning, $f(\infty)$ must exist. In the case of the geometric transform this means that all poles of $G(f(n))$ must lie outside the unit circle. For the z-transform this means that all poles of $Z(f(n))$ must lie inside the unit circle.

Other useful operating rules are summarized in Appendix C.

Inversion of Discrete Transforms

Finding the inverse of discrete transforms is similar to finding the inverse of Laplace transforms. Although inversion of both geometric and z-transforms can be accomplished by contour integration, we will opt for the use of basic theorems, definitions, and tables of transform pairs to find the inverse transforms.

Partial fraction expansion is as useful for discrete transforms as it was for Laplace transforms. Whenever a ratio of polynomials with a factorable denominator in the transform variable z appears, it can be written as a sum of separate terms more easily located in the tables. For example, consider the transform

$$G(f(n)) = \frac{4 - 12z}{2z^2 - 3z + 1} = \frac{4 - 12z}{(1 - 2z)(1 - z)} \tag{2-75}$$

By partial fractions

$$G(f(n)) = \frac{c_1}{1 - 2z} + \frac{c_2}{1 - z}$$

where

$$c_1 = \lim_{z \to 0.5} (1 - 2z)G(z) = -4 \qquad c_2 = \lim_{z \to 1} (1 - z)G(z) = 8$$

Therefore, in expanded form

$$G(f(n)) = \frac{-4}{1 - 2z} + \frac{8}{1 - z} \tag{2-76}$$

From the tables in Appendix C it follows that

$$f(n) = 8 - 4(2^n) \tag{2-77}$$

In addition to the use of contour integration, basic identities, and partial fractions, we have at our disposal the techniques of long division and Maclaurin series expansion of the transform. These last two tools are especially useful when it is only necessary to know the value of the function $f(n)$ for low-order terms.

Consider, for example, the z-transform

$$Z(f(n)) = \frac{z}{z - a} \tag{2-78}$$

where a is a constant. By long division this can be expressed in terms of a power series in z^{-1} as

$$\begin{array}{r|l} & 1 + az^{-1} + a^2z^{-2} + \cdots \\ \hline z - a & z \\ & \underline{z - a} \\ & \quad a \\ & \quad \underline{a - a^2z^{-1}} \\ & \qquad a^2z - 1 \\ & \qquad \underline{a^2z - 1 - a^3z^{-2}} - \cdots \end{array} \tag{2-79}$$

From the definition of the z-transform we know that the coefficient on z^{-n} is the value of the function $f(n)$. It therefore follows from the result of the long division that

$$f(0) = 0$$
$$f(1) = a$$
$$f(2) = a^2$$
$$\vdots$$
$$f(n) = a^n$$

The Maclaurin series, which is a Taylor-series expansion around the argument $z = 0$, can also be used to develop coefficients for appropriate powers of z. We can consider the geometric transform as a continuous function in z even though it is discrete in n. By the Maclaurin series it follows that a geometric transform can be written as

$$G(z) = \sum_{n=0}^{\infty} \frac{1}{n!} \frac{d^nG(z)}{dz^n}\bigg|_{z=0} z^n \tag{2-80}$$

For example, consider the geometric transform

$$G(f(n)) = \frac{z}{(1 - z)^2} \tag{2-81}$$

From the Maclaurin series

$$f(0) = G(z)\bigg|_{z=0} = 0$$

$$f(1) = \frac{d}{dz} G(z)\bigg|_{z=0} = (1 - z)^{-2} + 2z(1 - z)^{-3}\bigg|_{z=0} = 1$$

$$f(2) = \frac{1}{2} \frac{d^2}{dz^2} G(z)\bigg|_{z=0}$$

$$= \frac{1}{2} [2(1 - z)^{-3} + 2(1 - z)^{-3} + 6z(1 - z)^{-4}]\bigg|_{z=0} = 2$$

As can be verified from the tables, the general term is $f(n) = n$. The Maclaurin series expansion obviously suffers from computational difficulties for any but the lowest order terms of the simplest type of transforms.

To use the Maclaurin series for z-transforms, the approach is the same, only now the function must be expanded in powers of z^{-1}. This is accomplished by first rewriting the transform as a function of $r = z^{-1}$, then expanding in a Maclaurin series about $r = 0$. That is, let

$$R(r) = Z(z)\Big|_{z^{-1}=r}$$

Then

$$f(n) = \frac{1}{n!}\frac{d^nR}{dr^n}\Big|_{r=0}$$

For example, suppose that the z-transform of $f(n)$ is

$$Z(f(n)) = \frac{z(1+z)}{(z-1)^3} \tag{2-82}$$

By changing variables, we obtain

$$R(r) = \frac{r(r+1)}{(1-r)^3} \tag{2-83}$$

from which

$$f(0) = R(0) = 0$$

$$f(1) = \frac{dR}{dr}\Big|_{r=0} = 1$$

$$f(2) = \frac{1}{2}\frac{d^2R}{dr^2}\Big|_{r=0} = 4$$

$$\vdots$$

$$f(n) = \frac{1}{n!}\frac{d^nR}{dr^n}\Big|_{r=0} = n^2$$

Convolutions

From the point of view of the applications contained in the balance of this book, the most important operation in operational mathematics is that of finding the transform of the convolution of two functions. Briefly stated, the convolution of two functions is equal to the inverse of the product of their transforms. This theorem is sometimes referred to as Borel's theorem.

The relationship is applicable for all of the transform types, both continuous and discrete, discussed in this book. Symbolically,

$$T(f * g) = T(f)T(g) \tag{2-84}$$

where T represents any of our transform operators, f and g are the functions being convolved, and the star represents the convolution operation.

For discrete functions, the convolution $f * g$ of the functions $f(n)$ and $g(n)$ is defined as the summation

$$f * g = \sum_{i=0}^{n} f(i)g(n-i) \tag{2-85}$$

This function arises in probability modeling whenever it is necessary to find the distribution of the sum of two integer-valued independent random variables. It also arises in linear system analysis when g is the impulse response function and f any arbitrary driving function. (The transform of g is referred to as a transfer function in such a context.) In addition, this relationship can be used to invert transforms which can be broken into recognizable products of the transform variable.

To demonstrate the validity of Eq. (2-84) for a particular discrete transform, let us take the geometric transform of the convolution sum:

$$\begin{aligned} G(f * g) &= \sum_{n=0}^{\infty} \left(\sum_{i=0}^{n} f(i)g(n-i) \right) z^n \\ &= \sum_{n=0}^{\infty} \sum_{i=0}^{n} f(i)z^i g(n-i)z^{n-i} \\ &= \sum_{i=0}^{\infty} f(i)z^i \sum_{n=i}^{\infty} g(n-i)z^{n-i} \\ &= \sum_{i=0}^{\infty} f(i)z^i \sum_{j=0}^{\infty} g(j)z^j = G(f(n))G(g(n)) \end{aligned} \tag{2-86}$$

This very powerful result says that the convolution sum can be replaced by a simple product of geometric transforms. If the time function is required, the resulting transform must then be inverted. However, as we shall see in later chapters, we can often obtain much useful information about the convolution of two probability distributions directly from the transform without inverting.

The convolution operation for continuous functions is expressed as the integral

$$f * g = \int_0^t f(r)g(t-r)\, dr \tag{2-87}$$

This integral arises in probability modeling when it is necessary to find the distribution of the sum of two continuously distributed, independent random variables. It also arises in linear systems analysis and in the inversion of some transforms.

Let us use the Laplace transform to demonstrate the validity of Borel's theorem for continuous functions. Define $F(s) = \mathscr{L}(f(t))$ and $G(s) = \mathscr{L}(g(t))$. From the shifting properties of Laplace transforms we know that

$$e^{-sr}G(s) = \int_0^\infty e^{-st}g(t-r)U(t-r)\,dt = \mathscr{L}(g(t-r)U(t-r)) \quad (2\text{-}88)$$

This permits us to express the product of transforms as

$$\begin{aligned} F(s)G(s) &= \left(\int_0^\infty f(r)e^{-sr}\,dr\right)G(s) \\ &= \int_0^\infty f(r)e^{-sr}G(s)\,dr \\ &= \int_0^\infty f(r)\int_0^\infty e^{-st}g(t-r)U(t-r)\,dt\,dr \\ &= \int_0^\infty f(r)\int_r^\infty e^{-st}g(t-r)\,dt\,dr \end{aligned} \quad (2\text{-}89)$$

The region of integration is shown in Figure 2-6. If the first integration on t is changed to an integration on r first, where t ranges from r to infinity, r

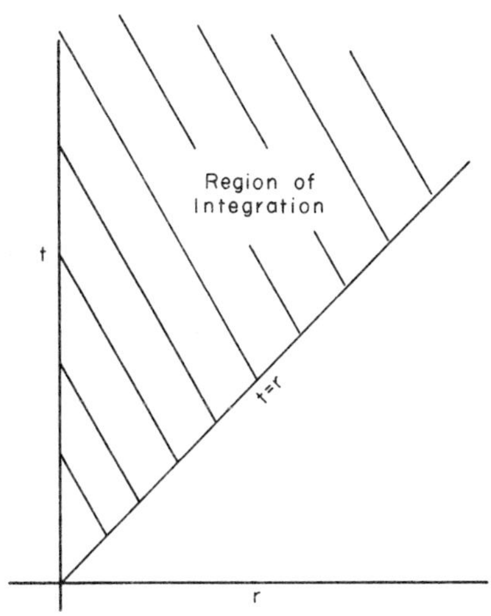

Figure 2-6 Region of integration.

ranges from zero to t. By changing the order of integration, we are then led to

$$F(s)G(s) = \int_0^\infty \int_r^\infty e^{-st} f(r) g(t - r)\, dt\, dr$$

$$= \int_0^\infty \int_0^t e^{-st} f(r) g(t - r)\, dr\, dt$$

$$= \int_0^\infty e^{-st} \int_0^t f(r) g(t - r)\, dr\, dt \tag{2-90}$$

But the inner integral is the expression for the convolution of f and g. It therefore follows that

$$F(s)G(s) = \mathscr{L}(f * g) \tag{2-91}$$

For a more rigorous analysis of this operation, see Churchill [4, pp. 43–49].

Before leaving the discussion of convolutions, it is well to note that the convolution operation is commutative, associative, and distributive. These properties will be useful in manipulating complex expressions in later exercises.

The commutative property implies that the order in which the functions being convolved are written is immaterial. That is,

$$f(t) * g(t) = g(t) * f(t) \tag{2-92}$$

The distributive property implies that

$$f(t) * [g(t) + h(t)] = f(t) * g(t) + f(t) * h(t) \tag{2-93}$$

The associative property implies that

$$f(t) * [g(t) * h(t)] = [f(t) * g(t)] * h(t) \tag{2-94}$$

We shall make extensive use of the convolution properties of transforms throughout the remainder of the book. Further details for such calculations will be introduced as required in specific problem contexts.

Some Properties of The Mellin Transform

From the point of view of our use of the Mellin transform the most important property is that of the Mellin convolution of two functions. This convolution is defined by the integral

$$f(x) = \int_0^\infty \frac{1}{y} g\left(\frac{x}{y}\right) h(y)\, dy \tag{2-95}$$

It happens that this is precisely the form of the probability density function of the random variable $x = ry$, where r and y are continuously distributed, independent random variables with probability density functions $g(r)$ and $h(y)$, respectively.

The transform of this special convolution reduces to a simple product of Mellin transforms in precisely the same manner that Fourier and Laplace transforms operated on the convolutions of the last section. If we define the Mellin transform of $f(x)$ as $F_{\mathrm{M}}(s)$, it follows that

$$F_{\mathrm{M}}(s) = G_{\mathrm{M}}(s)H_{\mathrm{M}}(s) \tag{2-96}$$

where $G_{\mathrm{M}}(s)$ and $H_{\mathrm{M}}(s)$ are the Mellin transforms of the functions g and h used in Eq. (2-95).

There is a complex inversion formula which enables one to recover the function $f(x)$ from the transform $F_{\mathrm{M}}(s)$. That formula is

$$f(x) = \frac{1}{2\pi i} \lim_{b\to\infty} \int_{c-ib}^{c+ib} x^{-s} F_{\mathrm{M}}(s)\, ds \tag{2-97}$$

where the path of integration is a line parallel to the imaginary axis and lying within the strip of analyticity of $F_{\mathrm{M}}(s)$ [12, p. 351].

Another important property of the Mellin transform is what Churchill refers to as the operational property [4, p. 453]. Defining

$$M_s(f(x)) = F_{\mathrm{M}}(s)$$

it can be shown that the operation of differentiation and multiplication by powers of x can be replaced by

$$M_s\left\{\left(x\frac{d}{dx}\right)^n f(x)\right\} = (-1)^n s^n F_{\mathrm{M}}(s) \tag{2-98}$$

Note the resemblance of this operational property to that of the Laplace transform of derivatives of functions. As a special case, it follows that

$$M_s(xF'(x)) = -sF_{\mathrm{M}}(s) \tag{2-99}$$

For higher order derivatives one must repeatedly apply the operator as in

$$M_s\left\{\left(x\frac{d}{dx}\right)^2 f(x)\right\} = M_s\left\{\left(x\frac{d}{dx}\right)(xf'(x))\right\} = M_s\{xf'(x) + x^2 f''(x)\}$$

from which

$$M_s(x^2 f''(x)) = (-1)^2 s^2 F_{\mathrm{M}}(s) - M_s(xf'(x)) = (s^2 + s)F_{\mathrm{M}}(s) \tag{2-100}$$

We shall return to other properties of the Mellin transform pertaining to probability functions in later chapters.

Problems

1. Expand the Fourier series representation of the function plotted in Figure 2-1 to include the first five nonzero terms of the series. Plot the resulting function, the original curve, and the approximation of Figure 2-1 on the same axes. Comment on the usefulness of higher order terms in such approximations.

2. Daily sales in the carry-out business seem to follow a periodic pattern. An industry analyst has had the course in mathematical methods and feels that he can predict the sales pattern. His forecasting model is

$$b(t) = k \cos\left[\frac{\pi}{12}(t - t_0)\right] + c$$

He has data for one period which he hopes to use to estimate the parameters k, c, and t_0. His criterion is to minimize the total square error over that period.

(a) What is the length of the period in this system?

(b) Find estimates for the parameters expressed as functions of his available data $a(t)$, $t = 0, 1, 2, \ldots$ [$b(t)$ is to be used to forecast $a(t)$].

3. The record's bureau sergeant for the campus police has collected some scattered data concerning daily arrests on campus. His informants are not always capable of remembering when they should report, so the data are not as complete as he had hoped. In any event, he believes that the underlying process is best described by a function of the form

$$f(t) = C_0 + C_1 \cos\left(\frac{\pi t}{8} - d_1\right)$$

He believes that he is faced with a periodic process where time is measured in days beginning with May 31 and with a fundamental period of 16 days. Find the parameters C_0, C_1, and d_1 that will provide the minimum square error based upon the following data (state all assumptions):

Date	Arrests
June 2	16
June 4	10
June 8	14
June 10	18
June 14	12
June 16	10

4. Consider the Fourier series expansion for a periodic function of period

T as follows:

$$f(t) = 400 + \sum_{\substack{n=-\infty \\ n \neq 0}}^{\infty} \frac{400i}{n\pi} \exp \frac{n\pi t i}{5} \qquad \text{where} \quad i = \sqrt{-1}$$

(a) What is the period of the function $f(t)$?

(b) Find the amplitude and phase angle, measured relative to a pure cosine wave, for the fourth harmonic.

(c) Write the third partial sum for this series in the standard sine and cosine format first introduced for Fourier series.

5. Consider the function

$$f(x) = x^2, \quad -\pi < x < \pi, \qquad \text{where} \quad f(x + 2\pi) = f(x)$$

Use this function and your knowledge of Fourier series to evaluate the sum

$$\sum_{n=1}^{\infty} \frac{1}{n^2}$$

Hint: Check the function at $x = \pi$.

6. Prove that the following relationships for Laplace transforms are valid:

(a) $\mathscr{L}(tg(t)) = -\frac{d}{ds} G(s)$ (b) $\mathscr{L}\left(\frac{1}{t} g(t)\right) = \int_s^{\infty} G(s)\, ds$

(c) $\mathscr{L}(g(at)) = \frac{1}{a} G\left(\frac{s}{a}\right)$ (d) $\mathscr{L}\left(\frac{e^{at} t^{n-1}}{\Gamma(n)}\right) = \frac{1}{(s-a)^n}$

(e) $\mathscr{L}(\sin\,(at)) = \frac{a}{s^2 + a^2}$

7. Use the relationship for the Laplace transform of the derivative of a function to prove that the final value theorem is true:

$$\lim_{s \to 0} sF(s) = \lim_{t \to \infty} f(t)$$

8. Find the Laplace transform of the cumulative distributions $\int_0^x f(x)\, dx$ for each of the following probability density functions:

(a) $f(x) = 5e^{-5x}, \quad x > 0$

(b) $f(x) = \frac{x^4 \exp\,(-0.5x)}{\Gamma(5) 2^5}, \quad x > 0$

(c) $f(x) = 0.05, \quad 10 \le x \le 30$

9. Find the inverse transform

$$\mathscr{L}^{-1}\left[\frac{1-e^{-s}}{s(1-re^{-s})}\right]$$

10. (a) Solve the following integral equation for $\mathscr{L}(y(t))$ expressed as a proper fraction in s:

$$\int_0^t y(u)(t^2 - 2tu + u^2 - 4t + 4u + 4)\,du = \sin(5nt)$$

(b) Without inverting, what can you say about the components of $y(t)$ as t goes to infinity? Explain your reasoning.

(c) Find the function $y(t)$. (You do not need to reduce the coefficients to lowest terms.)

11. Evaluate the following expressions:

(a) $\mathscr{L}[f(t)]$, where $f(t) = \begin{cases} \dfrac{1}{b-a}, & a \le t \le b \\ 0 & \text{elsewhere} \end{cases}$

(b) $\mathscr{L}^{-1}\left[\dfrac{6s}{s^4 + 18s^2 + 81}\right]$

(c) $\lim_{t\to\infty} f(t)$ when $\mathscr{L}[f(t)] = \dfrac{s^2 + 15 + 6s}{(s+6)(s^2+25)}$

12. Find the function $f(t)$ if $\mathscr{L}(f(t))$ equals

(a) $\dfrac{s+2}{s^2 + 4s + 4 + a^2}$ (b) $\dfrac{s^2 - a^2}{s^4 + 2s^2a^2 + a^4}$

(c) $\dfrac{s-1}{s^3 + s^2}$ (d) $\dfrac{2s}{s^4 + 2s^2 + 1}$

13. Prove that the following relationships for discrete transforms are valid:

(a) $G(k^{rn}g(n)) = G(g(n))\Big|_{z=k^rz}$ (b) $G(k^n) = \dfrac{1}{1-kz}$

(c) $Z(n) = \dfrac{z}{(z-1)^2}$ (d) $Z\left(\dfrac{k^n}{n!}\right) = e^{k/z}$

(e) $G(n^{(r)}) = \dfrac{r!\,z^r}{(1-z)^{r+1}}$

14. In order to expedite the calculation of Mellin transforms, it is sometimes convenient to use known Fourier transforms. Prove that the following relationship exists between the Fourier and Mellin transform for the function $f(x)$:

$$F_{\mathrm{M}}(f(x)) = \mathscr{F}(f(e^{-x})e^{-ax})$$

where a is the real part of the complex transform variable $s = a + ib$.

15. Develop the Mellin transforms for the following probability density functions:

(a) $f(x) = ae^{-ax}, \quad 0 < x < \infty$

(b) $f(x) = \dfrac{b(bx)^a e^{-bx}}{\Gamma(a+1)}, \quad 0 < x < \infty$

16. (a) An integral equation of the following form is often encountered in renewal theory:

$$h(t) = f(t) + \int_0^t h(t-u)f(u)\,du$$

where $h(t)\,dt$ is the probability density for a renewal in $(t, t + dt)$ and $f(t)$ is the pdf for time between events. Suppose it is known that $f(t) = 0.2 \exp(-0.2t)$. Using your knowledge of Laplace transforms, find the transform of the renewal density function $h(t)$ and the corresponding time function.

(b) After performing extensive calculations, the following transform has been obtained:

$$\mathscr{L}(y(t)) = \frac{5ns^3}{(s^2 + 25n^2)(4s^2 - 4s + 2)}$$

(i) Without inverting, what can you say about the components of $y(t)$ as t goes to infinity? Explain.

(ii) Find the function $y(t)$.

17. Consider the design of a pushcart to be used in the spring carnival races. The cart will be driven by a contingent of fraternity men. The designer is willing to assume that this driving force is a constant A. The driving force is being consumed by three factors: the acceleration imparted to the load (force = mass × acceleration), the wind resistance (force = c × velocity), and friction, which is independent of speed but increases with distance traveled. The friction is caused by foreign substances collecting at the bearing surfaces. The force consumed is $kx(t)$, where $x(t)$ measures distance traveled by time t.

(a) Develop a model to describe the performance of this system.

(b) Assuming that all contestants begin the race from a standstill, find how far the cart can be expected to go.

(c) The designer can influence the mass of the vehicle, the wind resistance, and the bearing design. If he wants to maximize distance traveled, which of the variables should receive his attention? Explain.

18. Consider the difference equation

$$f(n+2) - 5f(n+1) + 6f(n) = 0, \quad n \geq 0$$

where $f(0) = 0, f(1) = 1$. Find the function $f(n)$ that satisfies this equation.

19. Solve the differential–difference equation for $y_n(t)$ if

$$\frac{dy_{n+1}(t)}{dt} = y_n(t)$$

and

$$y_0(t) = 1, \quad t > 0, \qquad y_k(0) = 1, \quad k = 0, 1, 2, \ldots$$

20. Solve the following differential–difference equation for the function $f_n(t)$, where n is discrete and t is continuous:

$$\frac{df_n(t)}{dt} - 5f_{n-1}(t) + 5f_n(t) = 0$$

where $f_0(0) = 1$ and $f_n(0) = 0$ for $n \geq 1$.

CHAPTER 3

BASIC PROBABILITY IMPLICATIONS OF TRANSFORMS

The techniques of operational calculus are particularly useful in the context of analyzing probabilistic models of systems. Their power stems from the same observations made throughout our previous discussions. In the probability modeling context, it is often possible to reduce complex operations involving differentiation and integration to simple algebraic manipulations in the transform domain. In addition to the usual operations of simple differentiation and integration, transforms have the added advantage in probability models of providing convenient computational techniques for finding the moments of distributions.

Common Problems

In order to motivate interest in the application of transform techniques in probability modeling, consider the frequently encountered problem of finding the distribution of the sum of two independent random variables. For example, suppose that a separate maintenance department is being planned for a small manufacturing company. This new department will service both the foundry and machine shop divisions, who presently handle their own maintenance. Past data exist to indicate the daily relative frequency of emergency maintenance calls for each of the divisions. In establishing crew size requirements to meet a specified minimum level of service, it is necessary to find the distribution for the number of daily calls to be handled by the new maintenance department. This is a simple sum of random variables problem, which we can solve by the techniques of this chapter. In the next chapter we will also see how to expand this approach to include an estimate of total daily time demand if the distribution for the length of each maintenance task is known. The problem is then one of a random sum of random variables.

Let the number of daily maintenance calls for the machine shop be x_1 and the number for the foundry be x_2. Suppose x_1 is distributed by $f_1(x)$ and x_2 by $f_2(x)$, where x_1 and x_2 are independent discrete random variables. The problem is to find the distribution for total daily calls $g(y)$, where

$$y = x_1 + x_2$$

To illustrate the required calculation, suppose that x_1 and x_2 have the following discrete distributions:

x_1	$f_1(x)$	x_2	$f_2(x)$
0	0.1	0	0.3
1	0.2	1	0.5
2	0.3	2	0.2
3	0.4		

The distribution $g(y)$ can be developed by simply listing the mutually exclusive and collectively exhaustive ways that x_1 and x_2 can total to y. Defining $\Pr(y = a)$ as the probability that $y = a$, we are led to the following:

$$\Pr(y = 0) = \Pr(x_1 = 0 \text{ and } x_2 = 0)$$

$$\Pr(y = 1) = \Pr(x_1 = 0 \text{ and } x_2 = 1 \text{ or } x_1 = 1 \text{ and } x_2 = 0)$$

$$\vdots$$

$$\Pr(y = a) = \Pr(x_1 = 0 \text{ and } x_2 = a \text{ or } x_1 = 1 \text{ and } x_2 = a - 1 \cdots \text{ or } x_1 = a \text{ and } x_2 = 0)$$

Rewriting these statements in terms of the given probability distributions, we obtain

$$g(0) = f_1(0)f_2(0) = (0.1)(0.3) = 0.03$$

$$g(1) = f_1(0)f_2(1) + f_1(1)f_2(0) = (0.1)(0.5) + (0.2)(0.3) = 0.11$$

$$g(2) = f_1(0)f_2(2) + f_1(1)f_2(1) + f_1(2)f_2(0)$$
$$= (0.1)(0.2) + (0.2)(0.5) + (0.3)(0.3) = 0.21$$
$$g(3) = f_1(0)f_2(3) + \cdots + f_1(3)f_2(0)$$
$$= (0.1)(0) + (0.2)(0.2) + \cdots + (0.4)(0.3) = 0.31$$
$$g(4) = f_1(0)f_2(4) + \cdots + f_1(4)f_2(0) = 0.26$$
$$g(5) = f_1(3)f_2(2) = 0.08$$

The tedium of carefully enumerating all of the cases above is relieved somewhat by writing the general term as

$$g(y) = \sum_{i=0}^{y} f_1(i)f_2(y-i) \tag{3-1}$$

Equation (3-1) is the familiar convolution sum, which we encountered in our earlier discussion of transform techniques. On the basis of that discussion it seems reasonable to suppose that an appropriate transform for the $g(y)$ sequence could be developed from the transforms of the $f_1(x)$ and $f_2(x)$ sequences. By using Borel's theorem it follows that

$$T(g(y)) = T(f_1(x))T(f_2(x)) \tag{3-2}$$

where T denotes any appropriate discrete transform operator.

Since Eq. (3-1) is a perfectly general expression for the convolution of any two discrete functions f_1 and f_2, it follows that Eq. (3-2) can be applied whenever it is possible to find appropriate expressions for the transforms of f_1 and f_2. Recovery of the distribution $g(y)$ is accomplished by inverting the transform $T(g(y))$. Because the indicated inversion is often a very complicated task, much of our effort in this chapter will be devoted to developing methods to avoid the inversion process.

Once the arguments for finding the distribution of the sum of two independent random variables are understood, it is an easy matter to extend those arguments to find the distribution for the sum of n independent random variables. For example, suppose that the maintenance department of our previous example is to be expanded to include service for the assembly division with x_3 calls per day. Let the sum

$$y = x_1 + x_2 + x_3$$

be rewritten as

$$y = z + x_3, \qquad \text{where} \quad z = x_1 + x_2$$

From our previous discussion it follows that

$$h(z) = \sum_{i=0}^{z} f_1(i)f_2(z-i)$$

or

$$T(h(z)) = T(f_1(x))T(f_2(x))$$

Furthermore,

$$g(y) = \sum_{i=0}^{y} h(i) f_3(y - i) \qquad \text{and} \qquad T(g(y)) = T(h(z))T(f_3(x))$$

Remembering that the convolution operation is commutative, associative, and distributive, we see that

$$T(g(y)) = T(f_1(x))T(f_2(x))T(f_3(x))$$

It is obvious that the argument can be easily extended to the general case of n variables. Given

$$y = x_1 + x_2 + \cdots + x_n$$

it follows that

$$T(g(y)) = \prod_{j=1}^{n} T(f_j(x)) \tag{3-3}$$

where the following short-hand "product" notation is used:

$$\prod_{j=1}^{n} T_j = T_1 T_2 T_3 \cdots T_n$$

The results noted above for discrete functions are easily transferred to the case of continuously distributed random variables. Suppose the x_i are continuously distributed, independent random variables with probability density functions $f_i(x)$. The distribution of the sum

$$y = x_1 + x_2$$

can be expressed as the convolution integral

$$g(y) = \int_{-\infty}^{y} f_1(x) f_2(y - x)\, dx \tag{3-4}$$

Again applying Borel's theorem, it follows that

$$T(g(y)) = T(f_1(x))T(f_2(x)) \tag{3-5}$$

where the transforms are appropriately selected transforms for continuous functions.

The obvious extension for n variables

$$y = x_1 + x_2 + \cdots + x_n$$

leads to

$$T(g(y)) = \prod_{j=1}^{n} T(f_j(x)) \tag{3-6}$$

One of the more helpful extensions involving the use of transforms in the convolution of probability distributions occurs when the number of terms in the summation is itself permitted to be a random variable. We will refer to this as a random sum of random variables, about which we will have much more to say in later sections.

Once our modeling demands get beyond simple sums of random variables or random sums of random variables, the commonly encountered transforms will be of less immediate value. For those situations not easily addressed by transforms one must resort to approximation techniques for finding moments of distributions of compound random variables, such as those based upon the Taylor series, or to Monte Carlo experiments.

In order to circumvent some of the difficulties encountered in finding inverse transforms and indeed some of the difficulties in evaluating the moments of many well-known probability distributions, we will utilize certain mathematical properties of the various transforms. From such properties we will evaluate the moments of distributions directly from the transform rather than first inverting and then integrating. The balance of this chapter is devoted to investigating the properties of various kinds of transforms that are especially useful in the probability modeling context. In the process we hope to show that the general manipulative skills previously developed for transforms can all be put to good use in this new context.

Renaming the Transforms

It is unfortunate but true that practitioners in nearly every field of study attempt to associate new names with old tools when those tools are employed in their special field of interest. This practice is especially striking in the literature on probability and statistics. Terms such as "characteristic function" and "moment generating function" are used in most basic texts in the field without any hint that these are merely new names for some very old transforms.

The benefits to be gained from the application of what seem to be special-purpose transforms in probability contexts can be greatly expanded once the operators are recognized by their more proper names. For example, the characteristic function is nothing more than the inverse Fourier transform of a probability density function. Once it has been so identified the great number of available tables of transform pairs and theorems to assist in manipulations of the Fourier transform can be utilized. This insight can provide the systems analyst with a much higher success frequency in finding characteristic functions and their inverses than he may experience as a result of a specialized introduction to the topic.

The commonly employed transforms in probability modeling and their corresponding names from the more classical mathematical literature are listed in Table 3-1. Note that once a transform has been identified, e.g., the Laplace transform of $f(x)$, all of the theorems concerning that transform

Table 3-1

Renaming Transforms

Characteristic function	Fourier transform	$C(s) = \int_{-\infty}^{\infty} e^{isx} f(x)\, dx$
Moment generating function	Exponential transform	$M(s) = \int_{-\infty}^{\infty} e^{sx} f(x)\, dx$
	Laplace transform	$F(s) = \int_{0}^{\infty} e^{-sx} f(x)\, dx, \quad x \geq 0$
Generating function	Geometric transform	$G(z) = \sum_{i=0}^{\infty} z^i f(i), \quad i = 0, 1, 2, \ldots$

previously discussed can be employed. The fact that $f(x)$ may be a probability density function is incidental to the analysis. For our purposes it is a simple function of the variable x which we will transform into the frequency domain whenever that suits our needs.

The Characteristic Function

The characteristic function corresponding to a probability density function $f(x)$ is often defined as the expected value of e^{isx}, where i is the imaginary number $\sqrt{-1}$ and s is a transform variable:

$$C(s) = E(e^{isx}) = \int_{-\infty}^{\infty} e^{isx} f(x)\, dx \tag{3-7}$$

Sometimes this definition is modified slightly and rewritten in terms of the Stieltjes integral:

$$C(s) = \int_{-\infty}^{\infty} e^{isx}\, dF(x) \tag{3-8}$$

When expressed in this manner, the characteristic function is well defined for both continuous and discrete distributions. The term $F(x)$ represents the cumulative probability distribution. When x is a continuously distributed random variable, the term $dF(x)$ can be read as $f(x)\, dx$. In this instance, the Stieltjes integral reduces to the more familiar Riemann form of Eq. (3-1).

From our point of view the major advantage of the more general Stieltjes form of the integral is that it can permit us to develop general theorems concerning the manipulation of random variables without being concerned with whether the random variables involved are continuous or discrete. As a practical matter, we will use the characteristic function exclusively for continuous functions and employ only the Riemann form of the integral for the balance of this book. The reader should be aware of the alternate form, however, since it often appears in the more abstract treatises.

The s-term, which is recognizable as the frequency term in the Fourier transform, will be treated as a dummy variable in this chapter. Questions of existence of the characteristic function are easily answered by recalling the requirements for the Fourier integral. It has been previously noted that the Fourier integral, and consequently the Fourier transform, exists whenever the function being transformed has a finite number of points of discontinuity and the improper integral

$$\int_{-\infty}^{\infty} |f(x)|\, dx$$

exists. Since the function $f(x)$ is a probability density function, the indicated integral will always have a value of unity. The guaranteed existence of that improper integral for any probability density function is the basis for the statement that the characteristic function always exists.

With the question of existence so easily resolved, it is apparent why the characteristic function is held in such favor by theoreticians who want to use transforms to prove general theorems in probability theory. However, since it is somewhat more unwieldly than other transforms for numerical computations involving specific distributions, we will not make extensive use of the characteristic function in our calculations.

Recall from the general discussion of the Fourier transform that the transform is unique. This means that each probability density function has only one characteristic function and conversely that each characteristic function is associated with only one probability density function. Such a one-to-one relationship permits us to move back and forth from the time domain to the frequency domain with complete assurance that our calculations will not introduce spurious results.

Calculating Moments with the Characteristic Function

In addition to providing a powerful theoretical tool to prove relationships among distributions, the characteristic function also can be used to generate the moments of a probability distribution. The calculation of moments requires that we take derivatives with respect to the transform variable s and then evaluate the result for the point $s = 0$.

To illustrate, let us begin with the characteristic function of $f(x)$ itself evaluated at $s = 0$:

$$C(s)\bigg|_{s=0} = \int_{-\infty}^{\infty} e^{i0x} f(x)\,dx = \int_{-\infty}^{\infty} 1 \cdot f(x)\,dx = 1 \tag{3-9}$$

This relationship can often be used as a quick check on the algebra in developing a particular characteristic function. If the end result evaluated at $s = 0$ is not unity, the function present is obviously not a proper characteristic function.

By treating the transform variable s as a continuous variable, it is possible to take derivatives of $C(s)$ with respect to s. Since the variable s does not appear in the limits of the integral and is not the variable of integration, the differentiation all takes place under the integral sign. That is,

$$\begin{aligned}\frac{dC(s)}{ds}\bigg|_{s=0} &= \int_{-\infty}^{\infty} \frac{d}{ds}[e^{isx} f(x)]\,dx\bigg|_{s=0} \\ &= \int_{-\infty}^{\infty} ixe^{isx} f(x)\,dx\bigg|_{s=0} = i\int_{-\infty}^{\infty} xf(x)\,dx \end{aligned} \tag{3-10}$$

But the integral of Eq. (3-10) is nothing more than the expected value of the random variable x. It follows then that we have "generated" the first moment of the distribution, albeit multiplied by the imaginary number i, from the first derivative of the characteristic function. Rewriting, we have

$$E(x) = \frac{1}{i}\frac{dC(s)}{ds}\bigg|_{s=0} \tag{3-11}$$

Higher order moments follow directly from the higher order derivatives of $C(s)$. For example,

$$\frac{d^2C(s)}{ds^2} = \int_{-\infty}^{\infty} (ix)^2 e^{isx} f(x)\,dx$$

from which

$$E(x^2) = \frac{1}{i^2}\frac{d^2C(s)}{ds^2}\bigg|_{s=0} \tag{3-12}$$

In similar fashion, the rth moment about zero can be related to the rth derivative of $C(s)$ as

$$\frac{d^rC(s)}{ds^r} = \int_{-\infty}^{\infty} (ix)^r e^{isx} f(x)\,dx$$

or

$$E(x^r) = \frac{1}{i^r} \frac{d^r C(s)}{ds^r} \bigg|_{s=0} \tag{3-13}$$

Example It is often accepted by students of elementary probability and statistics that it is axiomatic that the sum of normally distributed random variables is normal, with the mean of the sum equal to the sum of the means and the variance of the sum equal the sum of the variances. The proof of this assertion is trivial when using characteristic functions.

Let x_1 be distributed by

$$f_1(x_1) = \frac{1}{(2\pi)^{1/2}\sigma_1} \exp \frac{-(x_1 - \mu_1)^2}{2\sigma_1^2}$$

and x_2 distributed by

$$f_2(x_2) = \frac{1}{(2\pi)^{1/2}\sigma_2} \exp \frac{-(x_2 - \mu_2)^2}{2\sigma_2^2}$$

The characteristic function associated with each of these distributions is of the form

$$C(s) = \frac{1}{(2\pi)^{1/2}\sigma} \int_{-\infty}^{\infty} \exp \left(\frac{2isx\sigma^2 - (x - \mu)^2}{2\sigma^2} \right) dx = \exp \left(i\mu s - \frac{\sigma^2 s^2}{2} \right) \tag{3-14}$$

If we define y as

$$y = x_1 + x_2$$

it follows from the convolution property of the characteristic function that

$$\begin{aligned} C_y(s) = C_1(s)C_2(s) &= \exp\,(i\mu_1 s - \tfrac{1}{2}\sigma_1^2 s^2) \exp\,(i\mu_2 s - \tfrac{1}{2}\sigma_2^2 s^2) \\ &= \exp\,[is(\mu_1 + \mu_2) - \tfrac{1}{2}s^2(\sigma_1^2 + \sigma_2^2)] \end{aligned} \tag{3-15}$$

When one compares the form of Eq. (3-15), with that of Eq. (3-14), it is obvious that the transforms have the same general appearance. The corresponding parameters are

$$\mu = \mu_1 + \mu_2\,, \qquad \sigma^2 = \sigma_1^2 + \sigma_2^2$$

Furthermore, since we know that the characteristic function is unique, it follows that the distribution of y must be normal with the indicated mean and variance.

In order to illustrate the calculation of moments, let us take the first two derivatives of Eq. (3-14). That is,

$$\frac{dC(s)}{ds} = (i\mu - s\sigma^2) \exp\left(i\mu s - \frac{\sigma^2 s}{2}\right)$$

and

$$\frac{d^2C(s)}{ds^2} = [-\sigma^2 + (i\mu - \sigma^2 s)^2] \exp\left(i\mu s - \frac{\sigma^2 s^2}{2}\right)$$

from which

$$\left.\frac{dC(s)}{ds}\right|_{s=0} = i\mu = iE(x)$$

and

$$\left.\frac{d^2C(s)}{ds^2}\right|_{s=0} = -\sigma^2 + (i\mu)^2 = i^2E(x^2)$$

It follows that

$$E(x) = \mu \qquad \text{and} \qquad \operatorname{var}(x) = E(x^2) - [E(x)]^2 = \sigma^2 + \mu^2 - \mu^2 = \sigma^2$$

as expected from our knowledge of the parameters of the normal distribution.

Moment Generating Functions

Moment generating functions receive their name from their principal use in probability and statistics, namely generating the moments of probability distributions.

The usual definition of the moment generating function is as the expected value of e^{sx}, where s is treated as a dummy variable. That is,

$$M(s) = E(e^{sx}) = \int_{-\infty}^{\infty} e^{sx} f(x)\, dx \tag{3-16}$$

is the moment generating function for the distribution $f(x)$. From our previous discussion, this is easily recognized as the exponential transform of the function $f(x)$. In most calculations involving the moment generating function, the transform variable s will be assumed to fall in the interval [0, 1]. Questions of convergence for various values of s outside that range will be ignored. If the indicated integral exists, we will use it.

For our purposes, $E(e^{sx})$ will be used only for continuously distributed random variables. When it exists, the moment generating function can be employed to perform the same tasks we might otherwise do with the characteristic function. However, two weaknesses arise in using the function $M(s)$. First, there is no guarantee that $M(s)$ exists for every continuous distribution. Second, $M(s)$ is an exponential transform and as such is not extensively tabulated. These weaknesses may cause us to ignore the moment generating function in favor of Fourier or Laplace transforms. One rarely encounters the exponential transform outside of the field of probability and statistics.

The Laplace transform, where defined, can be used as a moment generating function. In terms of expected values one could think of the Laplace transform as the expected value of e^{-sx}:

$$\mathscr{L}(f(x)) = E(e^{-sx}) = \int_0^{\infty} e^{-sx} f(x)\,dx, \qquad x \geq 0 \tag{3-17}$$

Note the close resemblance between the exponential and Laplace transforms. Since we will be using only the unilateral or one-sided Laplace transform, it is applicable only for those density functions defined over nonnegative x. Other than that restriction, it differs from the function $M(s)$ in the sign of the complex variable s. We will restrict our attention to the one-sided Laplace transform because it is much better tabulated than the bilateral Laplace transform and also because many operational problems use only nonnegative random variables.

The principal advantage of the Laplace transform over the exponential transform as a moment generating function is that it is much better known. Extensive tables and theorems exist for the Laplace transform, which can be used to assist our calculations. Knowledge of the exponential transform, on the other hand, is restricted to a very small number of special identities available in the probability literature.

Calculating Moments from Moment Generating Functions

Calculating the moments of a distribution by using the moment generating function involves computations quite similar to those used with the characteristic function.

Let us begin with the exponential transform $M(s)$, treated as a continuous function in s. As before, it is a useful check of one's calculations to evaluate any function that is assumed to be a moment generating function at the point $s = 0$. Since

$$M(s)\Big|_{s=0} = \int_{-\infty}^{\infty} e^{0x} f(x)\,dx = 1 \tag{3-18}$$

it follows that a function that does not yield a value of unity at $s = 0$ cannot be a moment generating function.

The moments about the origin for $f(x)$ can be recovered from $M(s)$ by taking an appropriate number of derivatives. The first moment is obtained from

$$\frac{d}{ds} M(s)\bigg|_{s=0} = \int_{-\infty}^{\infty} xe^{sx} f(x)\,dx\bigg|_{s=0} = E(x) \tag{3-19}$$

The obvious extension to higher order moments yields

$$E(x^r) = \frac{d^r}{ds^r} M(s)\bigg|_{s=0} = \int_{-\infty}^{\infty} x^r f(x)\,dx \tag{3-20}$$

In this manner it can be said that we have "generated" the moments of $f(x)$ from the function $M(s)$.

It is interesting to examine the function $M(s)$ expressed in terms of its Taylor series expansion about the point $s = a$:

$$\begin{aligned} M(s) &= M(a) + \frac{d}{ds} M(s)(s-a) + \frac{d^2}{ds^2} M(s)\frac{(s-a)^2}{2} + \cdots \\ &= \sum_{i=0}^{\infty} \frac{d^i}{ds^i} M(s)\frac{(s-a)^i}{i!} \end{aligned} \tag{3-21}$$

where the derivatives of $M(s)$ are evaluated at $s = a$. For the special case of $a = 0$, Eq. (3-21) reduces to a power series in s where the coefficient on the nth term is the nth moment of the distribution $f(x)$. That is,

$$M(s) = \sum_{i=0}^{\infty} \frac{d^i}{ds^i} M(s=0)\frac{s^i}{i!} = \sum_{i=0}^{\infty} E(x^i)\frac{s^i}{i!} \tag{3-22}$$

Since the moment generating function can be expressed as a sum of terms each multiplied by a moment of the distribution $f(x)$, one is faced with the intriguing possibility that if one could extract enough moments from sample data, he could use them to approximate $M(s)$ and in turn develop the probability density function $f(x)$.

The Laplace transform can also be employed in a manner similar to that used for the exponential transform to generate the moments of a distribution. Since

$$\mathscr{L}(f(x))\bigg|_{s=0} = \int_0^{\infty} e^{-0x} f(x)\,dx = 1 \tag{3-23}$$

it follows that a function that does not yield a value of unity at $s = 0$ cannot be a Laplace transform of any respectable probability distribution.

As before, the moments about the origin for the distribution $f(x)$ are recovered by taking an appropriate number of derivatives of the transform of $f(x)$. The only difference noted when using the Laplace transform to generate moments is that all odd-numbered derivatives will yield the negative of the corresponding moment. That is,

$$\left.\frac{d^r\mathscr{L}(f(x))}{ds^r}\right|_{s=0} = \int_0^\infty (-x)^r e^{-sx} f(x)\,dx \tag{3-24}$$

from which it is obvious that

$$E(x^r) = (-1)^r \left.\frac{d^r\mathscr{L}(f(x))}{ds^r}\right|_{s=0} \tag{3-25}$$

Example Consider a simple manual inspection system on a production line in which each unit being produced must undergo two separate final inspection tests by a single operator using two stations. The inspection time at station i follows the negative exponential distribution

$$f(t_i) = ae^{-at_i}, \qquad 0 \le t_i < \infty$$

Furthermore, there is no delay between stations since no new inspection is started until the current unit passes both tests. The problem is to analyze the total inspection time for a unit passing through the system, where

$$t = t_1 + t_2$$

The Laplace transform of the service time distribution at each station is given by

$$\mathscr{L}(f(t_i)) = \int_0^\infty ae^{-at_i}e^{-st_i}\,dt_i = \frac{a}{s+a}$$

If t_1 and t_2 are identically distributed, the application of the convolution theorem (3-2) yields

$$\mathscr{L}(f(t)) = [\mathscr{L}(f(t_i))]^2 = \left[\frac{a}{s+a}\right]^2$$

The expected time for a unit to clear the system is given by

$$E(t) = (-1)\left.\frac{d\mathscr{L}(f(t))}{ds}\right|_{s=0} = \left.\frac{2a^2}{(s+a)^3}\right|_{s=0} = \frac{2}{a}$$

The second moment for total time is

$$E(t^2) = \left.\frac{d^2\mathscr{L}(f(x))}{ds^2}\right|_{s=0} = \left.\frac{6a^2}{(s+a)^4}\right|_{s=0} = \frac{6}{a^2}$$

Combining terms, we can establish the variance as

$$\text{var}\,(t) = E(t^2) - [E(t)]^2 = \frac{6}{a^2} - \frac{4}{a^2} = \frac{2}{a^2}$$

The mean and variance calculations just performed are not very startling since we know that the mean of the exponential distribution is $1/a$ and the variance is $1/a^2$. Furthermore, the mean and variance of a sum of independent random variables are equal to the sum of the means and variances, respectively. The calculation does illustrate the technique, however, and provides the groundwork for later exercises in which simple addition is not sufficient.

This transform is one that is reasonably easy to invert. From the table of Laplace transform pairs we know that

$$\mathscr{L}(t^b) = \frac{\Gamma(b+1)}{s^{b+1}} \qquad \text{and} \qquad \mathscr{L}(e^{ct}f(t)) = \mathscr{L}(f(t))\Big|_{s=s-c}$$

It then follows that the total inspection time distribution is

$$\mathscr{L}^{-1}\left[\frac{a}{s+a}\right]^2 = \frac{a^2 te^{-at}}{\Gamma(2)} = f(t)$$

which can be recognized as a gamma distribution. What we have just shown is a special case of the general result that the sum of n exponentially distributed random variables is gamma-distributed. That is, if

$$t = t_1 + t_2 + \cdots + t_n$$

then

$$f(t) = \frac{a^n t^{n-1} e^{-at}}{\Gamma(n)}$$

Probability Generating Functions

The probability generating function commonly employed to analyze distributions of integer-valued random variables is nothing more than the geometric transform of the corresponding probability distributions. As such, we can employ all of the relationships developed earlier for manipulating the geometric transform of any sequence of real numbers.

Phrased in probability terms, one can think of the geometric transform of a probability distribution as the expected value of z^n where z is the complex transform variable and n is the integer-valued random variable distributed by $p(n)$. That is,

$$G(p(n)) = E(z^n) = \sum_{n=0}^{\infty} z^n p(n) \tag{3-26}$$

Although it is possible to consider negative-valued random variables in such a sum, we will limit our attention to those distributions defined only for positive n. Furthermore, we will restrict the use of the probability generating function to discrete distributions although it is possible to describe such a transform for continuous functions. The complex variable z will be treated as a dummy variable falling in the interval [0, 1] in our manipulations.

The name probability generating function is appropriate in the same sense that moment generating function is appropriate for Laplace transforms used in a probability context. One can "generate" the probability distribution by taking an appropriate number of derivatives of the geometric transform or probability generating function. Although the transform is discrete in the variable n, it can be treated as continuous in z and as such has derivatives.

In the transforms used so far in this chapter we have let the transform variable go to zero when we wanted numerical solutions for certain classes of generating problems. We will employ a similar strategy here. Since

$$G(p(n)) = \sum_{n=0}^{\infty} z^n p(n) = p(0) + p(1)z + p(2)z^2 + p(3)z^3 + \cdots$$

it follows that

$$G(p(n))\Big|_{z=0} = p(0)$$

Furthermore, by taking derivatives, we find

$$\frac{dGp(n)}{dz}\Big|_{z=0} = p(1) + 2p(2)z + 3p(2)z^2 + \cdots\Big|_{z=0} = p(1)$$

$$\frac{d^2G(p(n))}{dz^2}\Big|_{z=0} = 2p(2) + (3)(2)z + \cdots\Big|_{z=0} = 2p(2)$$

and in general

$$\frac{d^rG(p(n))}{dz^r}\Big|_{z=0} = (r!)p(r)$$

from which

$$p(r) = \frac{1}{r!}\frac{d^rG(p(n))}{dz^r}\Big|_{z=0} \tag{3-27}$$

What we have just demonstrated is that a probability distribution can be recovered from the power series expansion of its generating function. Expanding $G(p(n)) = G(z)$ in a Taylor series about the point $z = 0$ (the

Maclaurin series) yields

$$G(z) = G(0) + G'(0)z + \frac{1}{2}G''(0)z^2 + \cdots + \frac{1}{n!}G^n(0)z^{(n)} + \cdots$$

$$= \sum_{r=0}^{\infty} \left[\frac{d^r G(z)}{dz^r} \bigg|_{z=0} \right] \frac{z^r}{r!} \tag{3-28}$$

It then follows that $p(r)$ can be recovered from the coefficient of z^r in the Maclaurin series expansion of $G(z)$. Such an expansion may indeed be the easiest way to recover the probability distribution from certain classes of transforms with inverses that are hard to recognize. Even though a closed form for the entire distribution may not be easily recognized by this method, one can obtain numerical values for those frequency terms for which he has the patience to develop sufficiently higher order derivatives.

Calculating Moments from the Probability Generating Function

The probability generating function can also be used to "generate" the moments of a probability distribution in much the same manner that moment generating functions are used. From the definition, it follows that

$$G(p(n)) \bigg|_{z=1} = \sum_{n=0}^{\infty} p(n)z^n \bigg|_{z=1} = 1 \tag{3-29}$$

and we again have a quick check on the accuracy of our calculations. A function being proposed as a probability generating function is not a legitimate transform unless it yields a value of unity when $z = 1$.

If we now examine the derivatives of the transform at $z = 1$ instead of $z = 0$, as we did when attempting to generate the probability distribution, a new relationship emerges. For example,

$$\frac{dG(p(n))}{dz} \bigg|_{z=1} = \sum_{n=0}^{\infty} np(n)z^{n-1} \bigg|_{z=1} = E(n) \tag{3-30}$$

and we have recovered the first moment or mean of the distribution.

Continuing to higher derivatives gives a slightly different result than the one to which we are accustomed. That is,

$$\frac{d^2G(p(n))}{dz^2} \bigg|_{z=1} = \sum_{n=0}^{\infty} n(n-1)p(n)z^{n-2} \bigg|_{z=1} = E(n(n-1))$$

and in general

$$\frac{d^rG(p(n))}{dz^r} \bigg|_{z=1} = \sum_{n=0}^{\infty} n(n-1)\cdots(n-r+1)p(n)z^{n-r} \bigg|_{z=1}$$

$$= E(n(n-1)\cdots(n-r+1))$$

This expression takes on added significance if one recalls the definition of a factorial expression from the calculus of finite differences. Recall that $n^{(r)}$, which we read as n, r-factorial, can be written as

$$n^{(r)} = \frac{n!}{(n-r)!} = n(n-1)(n-2)\cdots(n-r+1)$$

It then follows that the derivatives of the geometric transform evaluated at $z = 1$ can be used to generate the factorial moments of the corresponding probability distribution. That is,

$$E(n^{(r)}) = \left.\frac{d^r G(p(n))}{dz^r}\right|_{z=1} \tag{3-31}$$

The resulting factorial moments can now be converted to corresponding power moments by using the Stirling number conversions introduced in Appendix A or by simple algebraic expansion for low-order moments. For example, since

$$n(n-1) = n^2 - n$$

it follows that

$$E(n^{(2)}) = E(n^2) - E(n)$$

Therefore, to develop an expression for the variance of a distribution as a function of its geometric transform, one uses

$$\text{var}\,(n) = E(n^2) - [E(n)]^2 = E(n^{(2)}) + E(n) - [E(n)]^2$$

from which

$$\text{var}\,(n) = G''(1) + G'(1) - [G'(1)]^2 \tag{3-32}$$

where

$$G'(1) = \left.\frac{dG(p(n))}{dz}\right|_{z=1}$$

Example A computer service center is considering expanding their service from strictly in-house operations to process jobs for both outside and in-house customers. The number of in-house customers is fixed at r, with each customer (department in this case) having a constant probability p of submitting a job in any given hour. For all practical purposes, the number of potential outside customers is infinite. Marketing personnel estimate that outside jobs will arrive at an average rate of a per hour. As an initial cut, the analyst is willing to describe the hourly outside customer demands by a Poisson distribution and the in-house customer demand by a binomial distribution. In order to justify additional equipment, it is necessary to develop statistics for the total job load.

This process is one in which a Poisson-distributed random variable n is added to a binomially distributed random variable b such that total jobs submitted is

$$k = n + b$$

The Poisson distribution with parameter a is given by

$$p(n) = \frac{a^n e^{-a}}{n!}$$

The geometric transform for this function is defined by

$$G(p(n)) = \sum_{n=0}^{\infty} \frac{a^n e^{-a} z^n}{n!} = e^{-a} \sum_{n=0}^{\infty} \frac{(az)^n}{n!} \tag{3-33}$$

But the infinite sum in this expression is simply the power series expansion for e^{az}. It therefore follows that

$$G(p(n)) = e^{-a} e^{az} = e^{-a(1-z)} \tag{3-34}$$

The binomial distribution with parameters r and p is given by

$$h(b) = \binom{r}{b} p^b (1-p)^{r-b}, \qquad 0 \le b \le r$$

Its geometric transform is, by definition,

$$G(h(b)) = \sum_{b=0}^{r} \binom{r}{b} p^b (1-p)^{r-b} z^b = \sum_{b=0}^{r} \binom{r}{b} (pz)^b (1-p)^{r-b} \tag{3-35}$$

A closed form expression for this sum is most easily developed by recalling the procedure for the binomial expansion in basic algebra. That is,

$$(a+b)^n = \sum_{i=0}^{n} \binom{n}{i} a^i b^{n-i} \tag{3-36}$$

By analogy with Eq. (3-35) it follows that

$$G(h(b)) = [pz + (1-p)]^r \tag{3-37}$$

Returning to the original problem of studying the sum of n plus b and using the convolution property of transforms, we have the transform for the distribution of k given as

$$G(f(k)) = G(p(n))G(h(b)) = e^{-a(1-z)}[pz + (1-p)]^r \tag{3-38}$$

The inversion of this transform is not immediately obvious. However, one could recover the first few terms of the distribution and the moments by taking derivatives. For example,

$$G(f(k))\bigg|_{z=0} = e^{-a}(1-p)^r = f(0)$$

and

$$G'(f(k)) = ae^{-a(1-z)}[pz + (1-p)]^r + rp[pz + (1-p)]^{r-1}e^{-a(1-z)}$$

from which

$$f(1) = G'(f(k))\bigg|_{z=0} = ae^{-a}(1-p)^r + rp(1-p)^{r-1}\, e^{-a}$$

and

$$E(k) = G'(f(k))\bigg|_{z=1} = a + rp$$

Higher order frequencies and moments are left as exercises for the reader.

Products and Quotients of Random Variables

Consider the problem faced by a vacation-bound aircraft owner in these times of uncertain fuel supplies. Planning the trip requires that the traveler make reservations at the resort area of his choice many weeks in advance. The pilot wants to limit consideration to those areas that he has a reasonably high probability of being able to reach without refueling, given recognized uncertainties in aircraft performance and fuel supplies. His fuel consumption in gallons per hour is a random variable influenced by pilot technique, engine conditions, and aircraft configuration. Ground speed is a random variable affected by power settings, aircraft trim condition, winds aloft, and required route deviations. Even the fuel available at time of takeoff is a random variable affected by local supplies, government regulations, and operator policies.

If every thing was deterministic, the proper equation defining our pilot's radius of action would be

$$d = gf/c \tag{3-39}$$

where d is the distance possible (miles), g the effective straight-line ground speed (miles per hour), f the fuel on board (gallons), and c the fuel consumption (gallons per hour). However, in this setting each of the variables listed is a random variable. If one assume that probability distributions are known for each variable separately, the problem then reduces to one of estimating the distribution for d. The tool we will use for analyzing problems of this type, which involve products and quotients of random variables, is the Mellin transform.

The Mellin Transform in Probability Modeling

The Mellin transform is useful in studying distributions for certain combinations of random variables for the same reasons that all of the other transforms discussed in this chapter are useful. One can replace often complicated integration arguments with simple algebraic operations on the transform. The key operation for products and quotients involves a special integral which we will call the Mellin convolution. This convolution can be operated on by Mellin transforms in much the same manner in which we operated on the convolution sums and integrals arising from the study of sums of random variables. For a more complete discussion of the material which follows, the reader should consult the paper by Epstein [12].

If x_1 and x_2 are independent random variables with continuous probability density functions $f_1(x)$ and $f_2(x)$, the density of the random variable $y = x_1 x_2$ can be obtained from

$$h(y) = \int_{-\infty}^{\infty} \frac{1}{|x|} f_1\left(\frac{y}{x}\right) f_2(x)\, dx = \int_{-\infty}^{\infty} \frac{1}{|x|} f_2\left(\frac{y}{x}\right) f_1(x)\, dx \tag{3-40}$$

If we further restrict our attention to only positive random variables, the resulting integral is in the form of a Mellin convolution of $f_1(x)$ and $f_2(x)$:

$$h(y) = \int_{0}^{\infty} \frac{1}{x} f_1\left(\frac{y}{x}\right) f_2(x)\, dx \tag{3-41}$$

From the discussion in Chapter 2, recall that the Mellin transform of a function $f(x)$ is defined as

$$F_{\mathrm{M}x}(s) = \int_{0}^{\infty} f(x) x^{s-1}\, dx \tag{3-42}$$

Furthermore, the transform of the Mellin convolution is given by

$$F_{\mathrm{M}}(h(y)) = F_{\mathrm{M}}(f_1(x)) F_{\mathrm{M}}(f_2(x)) \tag{3-43}$$

and the inverse transform is

$$h(y) = \frac{1}{2\pi i} \lim_{b \to \infty} \int_{c-ib}^{c+ib} x^{-s} F_{\mathrm{M}}(s)\, ds \tag{3-44}$$

Since we are mostly concerned with transforms of probability density functions, it is convenient to think of the Mellin transform in terms of expected values. Recall that the expected value of any function $g(x)$ of the random variable x, whose distribution is $f(x)$, is given by

$$E(g(x)) = \int_{-\infty}^{\infty} g(x)f(x)\,dx \tag{3-45}$$

It therefore follows that the Mellin transform of the density function $f(x)$ could be viewed as

$$F_{\mathrm{M}x}(s) = E(x^{s-1}) = \int_{0}^{\infty} x^{s-1}f(x)\,dx \tag{3-46}$$

The relationship of Mellin transforms to expected values makes it a simple matter to establish some important operating properties of the transform in probability modeling.

First, one may be faced with finding the distribution of some multiple of the random variable x, say $y = ax$. In terms of expected values, it follows that the Mellin transform of the distribution for y is given by

$$F_{\mathrm{M}y}(s) = E(y^{s-1}) = E((ax)^{s-1}) = a^{s-1}E(x^{s-1}) = a^{s-1}F_{\mathrm{M}x}(s) \tag{3-47}$$

The transform for $h(y)$ is obtained by multiplying the transform for x by a^{s-1}. For example, if x is uniformly distributed in the interval [0, 1], the transform of $f(x)$ is

$$F_{\mathrm{M}x}(s) = \int_{0}^{1} x^{s-1}\,dx = \frac{1}{s}$$

Then if $y = 15x$, it follows that the transform for $h(y)$ is

$$F_{\mathrm{M}y}(s) = \frac{15^{s-1}}{s}$$

A second calculation of importance is that of finding the distribution of powers of the random variable x. If $y = x^b$, the expected value argument leads to

$$\begin{aligned} F_{\mathrm{M}y}(s) &= E(y^{s-1}) = E((x^b)^{s-1}) \\ &= E(x^{s-1})\,\big|_{s=bs-b+1} = F_{\mathrm{M}x}(s)\,\big|_{s=bs-b+1} \end{aligned} \tag{3-48}$$

The transform for the distribution of the bth power of x can be obtained by replacing the s-argument in the transform of $f(x)$ by the term $(bs - b + 1)$. For example, if $y = x^3$, where x is uniformly distributed on [0, 1], it follows that

$$F_{\mathrm{M}y}(s) = \frac{1}{s}\bigg|_{s=3s-3+1} = \frac{1}{3s-2}$$

Equation (3-48) can also be used to find the transform of the inverse of a random variable. The variable $y = 1/x$ can be viewed as $y = x^b$ where $b = -1$. It therefore follows by direct substitution that

$$F_{My}(s) = F_{Mx}(s)\,|_{s=2-s} \tag{3-49}$$

The transform for the distribution of the inverse of the random variable x can be obtained by replacing the s-argument in the Mellin transform for $f(x)$ by the term $(2 - s)$. For example, suppose $y = 1/x$, where x is exponentially distributed with a mean of one. Then

$$f(x) = e^{-x}$$

from which

$$F_{Mx}(s) = \int_0^\infty x^{s-1}e^{-x}\,dx = \Gamma(s)$$

The transform for the distribution of y is then given by

$$F_{My}(s) = F_{Mx}(2 - s) = \Gamma(2 - s)$$

We are now in a position to develop an easy method for finding the transform of the distribution of a quotient of two random variables. Treat the random variable $y = x_1/x_2$ as a product of two independent random variables x_1 and $(1/x_2)$. By combining the results of Eq. (3-43) and (3-49), it follows that

$$F_{M(x_1/x_2)}(s) = F_{Mx_1}(s)F_{Mx_2}(2 - s) \tag{3-50}$$

For example, if x_1 and x_2 are both uniformly distributed on the interval [0, 1], the transform of the ratio $y = x_1/x_2$ is given by

$$F_{My}(s) = \frac{1}{s(2 - s)}$$

We are now ready to write the transform for the radius-of-action problem posed in Eq. (3-39). Since $d = gf/c$ can be viewed as a product of g and the quotient f/c, application of Eqs. (3-43) and (3-50) reveals that

$$F_{Md}(s) = F_{Mg}(s)F_{Mf}(s)F_{Mc}(2 - s) \tag{3-51}$$

The techniques for finding Mellin transforms of the probability density functions for products, powers, and quotients of independent random variables are summarized in Table 3-2. Mellin transforms for some particular density functions are given in Table 3-3. For a discussion of special techniques to handle negative-valued random variables, the reader should consult the paper by Epstein [12, pp. 374–375].

Table 3-2

Mellin Transform Operations for Products and Quotients

Random variable y	Mellin transform $F_{My}(s)$
x	$F_{Mx}(s) = \int_0^\infty x^{s-1} f(x)\,dx$
ax	$a^{s-1} F_{Mx}(s)$
x^b	$F_{Mx}(bs - b + 1)$
$1/x$	$F_{Mx}(2 - s)$
$x_1 x_2$	$F_{Mx_1}(s) F_{Mx_2}(s)$
x_1/x_2	$F_{Mx_1}(s) F_{Mx_2}(2 - s)$
$a x_1^{b_1} x_2^{b_2}$	$a^{s-1} F_{Mx_1}(b_1 s - b_1 + 1) F_{Mx_2}(b_2 s - b_2 + 1)$

Table 3-3

Mellin Transforms for Selected Densities

Density $f(x)$	Mellin transform $F_{Mx}(s)$
$f(x), \quad x > 0$	$\int_0^\infty x^{s-1} f(x)\,dx$
$1, \quad 0 \leq x \leq 1$	$\frac{1}{s}$
$\frac{1}{b - a}, \quad a \leq x \leq b$	$\frac{b^s - a^s}{s(b - a)}$
$e^{-x}, \quad x > 0$	$\Gamma(s)$
$ae^{-ax}, \quad x > 0$	$\left(\frac{1}{a}\right)^{s-1} \Gamma(s)$
$\frac{x^a e^{-x}}{\Gamma(a + 1)}$, $0 < x < \infty; \quad a > -1$	$\frac{\Gamma(a + s)}{\Gamma(a + 1)}$
$\frac{b(xb)^a e^{-bx}}{\Gamma(a + 1)}$, $0 < x < \infty; \quad a > -1; \quad b > 0$	$\left(\frac{1}{b}\right)^{s-1} \frac{\Gamma(a + s)}{\Gamma(a + 1)}$
$\frac{\Gamma(a + b)}{\Gamma(a)\Gamma(b)} x^{a-1}(1 - x)^{b-1}$, $0 < x < 1; \quad a > 0; \quad b > 0$	$\frac{\Gamma(a + b)\Gamma(a + s - 1)}{\Gamma(a + b + s - 1)\Gamma(a)}$

Calculating Moments from the Mellin Transform

Just as one can obtain moments directly from the Fourier, Laplace, and geometric transforms, so can one obtain moments from the Mellin transform without going through the inversion process. However, in this instance it is not even necessary to take derivatives. All that is necessary is to replace the s-argument in the transform with an appropriate constant.

From the definition

$$F_{Mx}(s) = \int_0^\infty x^{s-1} f(x)\, dx$$

it follows that replacing s by unity should give

$$F_{Mx}(1) = \int_0^\infty f(x)\, dx = 1 \tag{3-52}$$

This provides a check on one's algebra at the end of sometimes lengthy calculations. If the resulting transform evaluated at $s = 1$ does not yield unity, an algebraic mistake may be present.

Replacing s by 2 yields

$$F_{Mx}(2) = \int_0^\infty x f(x)\, dx = E(x) \tag{3-53}$$

In other words, the expected value of x is simply the Mellin transform of $f(x)$ evaluated at $s = 2$. Continuing in similar fashion, it follows that

$$E(x^r) = F_{Mx}(r + 1) \tag{3-54}$$

The rth moment about zero can be obtained by replacing the s-argument of the Mellin transform by $(r + 1)$.

By using the computational form for variance, it is obvious that

$$\text{var}\,(x) = E(x^2) - [E(x)]^2 = F_{Mx}(3) - [F_{Mx}(2)]^2 \tag{3-55}$$

With these results, we are now in a position to evaluate the mean and variance for various powers, products, and quotients of random variables directly from functions of their transforms. There is one caveat to be raised, however. The resulting transform may not be analytic for all values of s, especially when quotients are involved. If the transform is not analytic at $s = 2$ or $s = 3$, other techniques may be required to find the mean and variance.

Example Suppose $r = pq$ is a product of two random variables both uniformly distributed but each with a unique range.

$$f_p(x) = 0.025, \qquad 130 \le x \le 170$$

$$f_q(x) = 0.05, \qquad 40 \le x \le 60$$

The Mellin transform for a uniformly distributed random variable over the interval $[a, b]$ is

$$F_{\mathrm{M}x}(s) = \int_a^b \frac{x^{s-1}}{b-a}\,dx = \frac{b^s - a^s}{s(b-a)} \tag{3-56}$$

To check the validity of this result, let us evaluate the moments of the distribution using the transform. From Eqs. (3-54) and (3-55)

$$E(x^0) = F_{\mathrm{M}x}(1) = \frac{b-a}{b-a} = 1,$$

$$E(x) = F_{\mathrm{M}x}(2) = \frac{b^2 - a^2}{2(b-a)} = \frac{b+a}{2}$$

$$\mathrm{var}\,(x) = F_{\mathrm{M}x}(3) - [F_{\mathrm{M}x}(2)]^2 = \frac{b^3 - a^3}{3(b-a)} - \frac{1}{4}(b+a)^2 = \frac{(b-a)^2}{12}$$

All the above are consistent with the known values of mean and variance for the uniform distribution.

Applying these results to our problem, we have

$$F_{\mathrm{M}p}(s) = \frac{(170)^s - (130)^s}{40s}, \qquad F_{\mathrm{M}q}(s) = \frac{60^s - 40^s}{20s}$$

from which

$$F_{\mathrm{M}r}(s) = \frac{(170^s - 130^s)(60^s - 40^s)}{40(20)s^2}$$

The mean and variance for r are given by

$$E(r) = F_{\mathrm{M}r}(2) = \frac{(170^2 - 130^2)(60^2 - 40^2)}{(40)(20)(4)} = 7500$$

$$\begin{aligned}\mathrm{var}\,(r) &= F_{\mathrm{M}r}(3) - [F_{\mathrm{M}r}(2)]^2 \\ &= \frac{(170^3 - 130^3)(60^3 - 40^3)}{(40)(20)(9)} - (7500)^2 = 1{,}087{,}777\end{aligned}$$

Summary

The purpose of this chapter has been to demonstrate the basic calculations in probability modeling that can be facilitated by the use of transforms. We have demonstrated the use of Fourier, Laplace, exponential, geometric, and Mellin transforms as tools to find distributions of sums, products, and quotients of random variables and the moments of these distributions.

Table 3-4

Common Probability Distributions and Their Transforms

Distribution of random variable x	Formula	Range of x	Type of transform	Transform of $f(x)$
Continuous				
1 Normal	$(2\sigma^2\pi)^{-\frac{1}{2}} \exp[-(x-u)^2/2\sigma^2]$	$-\infty < x < \infty$	Fourier	$\exp(ius - \frac{1}{2}\sigma^2 s^2)$
2 Exponential	ae^{-ax}	$x \geq 0$	Laplace	$a/(s+a)$
3 Gamma	$x^{a-1}e^{-x/b}/\Gamma(a)b^a$	$0 \leq x < \infty$	Laplace	$[b^{-1}/(s+b^{-1})]^a$
4 Rectangular	$(b-a)^{-1}$	$a \leq x \leq b$	Laplace	$[1/s(b-a)](e^{-as} - e^{-bs})$
Discrete				
5 Binomial	$\binom{n}{x}p^x(1-p)^{n-x}$	$x = 0, 1, 2, \ldots, n$	Geometric	$[pz + (1-p)]^n$
6 Poisson	$u^x e^{-u}/x!$	$x = 0, 1, 2, \ldots, \infty$	Geometric	$e^{-u(1-z)}$
7 Geometric	$p(1-p)^x$	$x = 0, 1, 2, \ldots, \infty$	Geometric	$p/[1-(1-p)z]$

In order to provide a ready reference for future calculations, appropriate transforms for some of the better known probability distributions are listed in Tables 3-3 and 3-4. Some of these relationships have already been demonstrated in this chapter. Proofs that the remaining identities are valid are left as exercises for the reader.

The next chapter will build on the basic concepts introduced here to develop useful relationships for more complex modeling situations.

Problems

1. Find three applications from your personal experience in which models involving sums of random variables would be appropriate.

(a) Describe the problems.

(b) Discuss the data collection and curve fitting problems which you would have to face to model these situations.

(c) Suggest alternative means of solving your problems.

2. Develop the characteristic function for each of the following distributions:

(a) Exponential. (b) Gamma. (c) Rectangular.

3. Use the characteristic functions developed in Problem 2 to prove the general equations for the mean and variance of each of the indicated distributions.

4. Use your knowledge of transforms to find the appropriate distribution for

$$y = x_1 + x_2 + \cdots + x_n$$

where the x_i are identically distributed Poisson variables.

5. Use your knowledge of transforms to find the appropriate distribution for

$$y = x_1 + x_2 + \cdots + x_n$$

where the x_i are identically distributed binomial variables.

6. Use your knowledge of transforms to find the appropriate distribution for

$$y = x_1 + x_2 + \cdots + x_n$$

where the x_i are identically distributed exponential variables.

7. Use your knowledge of transforms to find the appropriate distribution for

$$y = x_1 + x_2 + \cdots + x_n$$

where the x_i are identically distributed geometric variables.

8. Modify the random variables in Problems 4 and 5 so that each x_i in a particular sum follows the same general family of distributions but has unique parameters. Find the appropriate distributions for the indicated sums.

9. Modify the random variables in Problems 6 and 7 so that each x_i in a particular sum follows the same general family of distributions but has unique parameters. Find the appropriate distributions for the indicated sums.

10. In the analysis of the local barber shop, it has been determined that the number of customers present at any one time can be described by the probability mass function p_n, where

$$3p_0 = 4p_1, \qquad 7p_n = 3p_{n-1} + 4p_{n+1}, \qquad n \geq 1$$

Use your knowledge of transforms to solve the following problems:

(a) Find the geometric transform of the distribution $\{p_n\}$.

(b) Find the geometric transform of the complementary cumulative distribution $P(k) = \Pr[n > k]$.

(c) Find the expected number of customers present.

(d) Find the variance for number of customers present.

11. In solving a certain stochastic process problem, the following function with a transform variable r has been obtained:

$$f(r) = \frac{0.2}{1 - 0.8r}$$

(a) Identify the type of transform.

(b) Prove whether or not it is the transform of a legitimate probability distribution.

(c) Use the transform to find the mean and variance of the distribution.

12. Find the geometric transform of the distribution $f(n)$ when it is known that n and t are random variables and that

$$G(f(n \mid t)) = \exp[-5t(1 - z)] \qquad \text{and} \qquad \mathscr{L}(g(t)) = \frac{1}{1 + 0.1s}$$

13. Consider the sequence $\{p_j\}$, where, p_j the probability that a random variable n takes the value j, and the associated transform $G(p_n) = \sum_{j=0}^{\infty} p_j z^j$ are known. Find a compact expression for the transform of the related sequence $\{q_j\}$, where q_j is the probability that $n > j$, expressed as a function of $G(p_n)$.

14. Consider a function of n variables

$$z = f(x_1, x_2, \ldots, x_n)$$

where each of the variables x_i has a known mean and variance. Use the Taylor series expansion for functions of n variables about the point $(a_1, a_2, \ldots, a_n)$

$$f(x_1, x_2, \ldots, x_n) = f(a_1, a_2, \ldots, a_n) + \sum_{i=1}^{n} (x_i - a_i) f_i(a_1, a_2, \ldots, a_n) + \frac{1}{2} \sum_{i=1}^{n} \sum_{j=1}^{n} (x_i - a_i)(x_j - a_j) f_{ij}(a_1, a_2, \ldots, a_n) + R$$

where

$$f_i = \frac{\partial f}{\partial x_i}, \qquad f_{ij} = \frac{\partial^2 f}{\partial x_i \, \partial x_j}$$

(a) If it is approximately true that

$$E(z) = f(u_1, u_2, \ldots, u_n)$$

prove that the variance for z can be approximated by

$$\operatorname{var}(z) = \sum_{i=1}^{n} \sum_{j=1}^{n} \left(\frac{\partial z}{\partial x_i}\right)\left(\frac{\partial z}{\partial x_j}\right) \sigma_{ij}$$

where all partials are evaluated at $x_i = u_i$ and σ_{ij} is the covariance for i and j.

(b) Use the results of Part (a) to prove that an approximate expression for the variance of a product of two random variables is

$$\operatorname{var}(x_1 x_2) \approx \mu_1^2 \sigma_2^2 + \mu_2^2 \sigma_1^2 + 2\mu_1 \mu_2 \sigma_{12}$$

(c) Prove that an approximate expression for the quotient of two random variables is

$$\operatorname{var}\left(\frac{x_1}{x_2}\right) \approx \frac{\sigma_1^2}{\mu_2^2} + \frac{\mu_1^2 \sigma_2^2}{\mu_2^4} - \frac{2\mu_1 \sigma_{12}}{\mu_2^3}$$

15. Compare the results of Problem 14 with values for the variance of products and quotients when the Mellin transform is used.

(a) Assume x_1 and x_2 are exponentially distributed with means of 2 and 4, respectively.

(b) Assume that x_1 and x_2 follow the rectangular distribution with (a, b) parameters of (2, 12) and (4, 8), respectively.

16. In the analysis of inventory systems a commonly recurring model is the simple deterministic equation

$$K = \frac{\lambda A}{Q} + \frac{ICQ}{2}$$

where λ is the annual demand rate, A the order cost, Q the order quantity, C the cost of each item, I the inventory carrying charge, and K the average annual cost.

(a) Suppose that all variables except Q are random variables for which the analyst has estimates of their means, variances, and covariances. Use the results of Problem 14 to estimate the mean and variance of K.

(b) Assume independence and the following density functions:

$$f_1(\lambda) = \frac{1}{400}, \qquad 800 \le \lambda \le 1200$$

$$f_2(A) = \frac{A \exp(-0.1A)}{100}, \qquad A \ge 0$$

$$f_3(I) = \frac{\Gamma(10)}{\Gamma(4)\Gamma(6)} I^3(1 - I)^5, \qquad 0 \le I \le 1$$

$$f_4(C) = 0.05, \qquad 40 \le C \le 60$$

(i) Find the Mellin transform for each of the densities.
(ii) Find the mean and variance of K by using the resulting transforms.

17. A taxi company operates a fleet of 15 taxicabs, five each from the three major automobile manufacturers. The company's past experience is that each taxi of brand A has a chance of 0.05 of being involved in an accident on any given day. The probabilities for brands B and C are 0.04 and 0.06, respectively.

(a) Find an appropriate transform for the probability distribution of total accidents expected by the company in any one day.

(b) Use the transform to find
(i) The probability of two or more accidents.
(ii) The expected number of accidents.
(iii) The variance for number of accidents.

18. Ice cream sales at the local swimming pool seem to increase throughout the week. The manager believes that demand on day i, where $i = 1$ implies Monday, averages

$$d_i = 300 + 20i$$

The range of daily demand seems to fall within 10% of the mean.
(a) Construct a model for weekly demand.
(b) Find an appropriate transform for the weekly demand distribution.
(c) Estimate the variance for weekly demand.

19. An automobile dealer is heavily committed to a sales incentive contest. He now has 12 cars on hand. The contest ends when his stock is exhausted.

He feels that it is reasonable to believe that the time between individual sales will be exponentially distributed with a mean of 0.5 day.

(a) Find the Laplace transform for the distribution of the length of the sales contest.

(b) Invert the transform in part (a).

(c) Find the variance for the contest duration.

(d) Find the probability that the contest will end in less than 5 days.

20. A sheet metal fabricating company has placed a purchase order for a new punch press. The machine tool manufacturer has told the company that their order is number 20 to be released to the shop for construction. The order queue is reduced in exponential fashion at a rate of one every 5 days. Once construction begins, the actual construction time for their press is estimated to be uniformly distributed between 20 and 30 days.

(a) Find a transform for the distribution of elapsed time until the company's order is released to the shop.

(b) Find a transform for the distribution of time it will take the company to receive their press.

(c) Find the mean and variance for delivery time.

CHAPTER 4

SIMPLE FUNCTIONS OF RANDOM VARIABLES

In the previous chapter we developed the basic techniques for using transforms in probability modeling. There we learned how to find the moments of distributions from each of several transform types. We also learned how to take advantage of the convolution properties of transforms to develop transforms for the distributions of sums, products, and quotients of independent random variables.

The techniques we discussed at length in Chapter 3 are of use in their own right to prove basic theorems in probability theory (e.g., show that the sum of Poisson variables is Poisson) and to calculate the moments of distributions with known transforms when the usual path of integration or summation appears intractable. However, these basic tools gain their real power when we begin to use them to evaluate distributions for random sums of random variables, solve infinite sets of differential–difference equations related to probability distributions, and examine other interesting applications encountered in model building. It is the purpose of this chapter to explore some of the more useful extensions of transforms in basic probability modeling, most of which simply involve repeated application of the basic convolution and moment generating properties of transforms.

Transforms for Functions of a Random Variable

We have already explored some simple cases for finding the transform of the distribution for functions of random variables. Consider again the sum of random variables where the x's are continuously distributed on the interval $(0, \infty)$:

$$y = x_1 + x_2 + \cdots + x_n \tag{4-1}$$

The variable y is a linear function of the variables x_1 through x_n. When the x variables are independent, we used Borel's theorem to establish

$$\mathscr{L}[f(y)] = E[e^{-ys}] = \mathscr{L}[f_1(x)]\mathscr{L}[f_2(x)] \cdots \mathscr{L}[f_n(x)] \tag{4-2}$$

What we have done of course is once again assert that the transform for the distribution of a sum of independent random variables can be written as the product of the transforms of the distributions of the random variables contained in that sum. The validity of Eq. (4-2) can be easily demonstrated with expected value arguments by noting that the joint density of independent random variables can be written as the product of the marginal densities. That is,

$$f(x_1, x_2, \ldots, x_n) = f_1(x)f_2(x) \cdots f_n(x) \tag{4-3}$$

Since $h(x_1, x_2, \ldots, x_n) = e^{-sy}$ is a function of the x variables, it follows from basic probability theory that

$$\begin{aligned} &E(h(x_1, x_2, \ldots, x_n)) \\ &\quad = \int_{x_1}\int_{x_2} \cdots \int_{x_n} h(x_1, x_2, \ldots, x_n) f(x_1, x_2, \ldots, x_n)\, dx_1 \cdots dx_n \\ &\quad = \int_{x_1}\int_{x_2} \cdots \int_{x_n} e^{-x_1 s} e^{-x_2 s} \cdots e^{-x_n s} f_1(x) f_2(x) \cdots f_n(x)\, dx_1 \cdots dx_n \end{aligned} \tag{4-4}$$

Regrouping terms, we obtain

$$\begin{aligned} &E(h(x_1, \ldots, x_n)) \\ &\quad = \int_{x_1} e^{-x_1 s} f_1(x)\, dx_1 \int_{x_2} e^{-x_2 s} f_2(x)\, dx_2 \cdots \int_{x_n} e^{-x_n s} f_n(x)\, dx_n \\ &\quad = \mathscr{L}[f_1(x)]\mathscr{L}[f_2(x)] \cdots \mathscr{L}[f_n(x)] \end{aligned} \tag{4-5}$$

as we asserted.

The next step is to employ the strategy demonstrated for the simple linear function for more complex functions of random variables. Let $\mathbf{X}$ be a vector with components x_i and let the joint density of the x_i values be given by $f(\mathbf{X})$. Treat any real-valued function of the variables x_i as a function $y(\mathbf{X})$. It follows that e^{-sy} is also a function of the variables x_i, say $h(\mathbf{X})$. We can then assert that

$$\mathscr{L}(f(y)) = E[\exp(-sy(\mathbf{X}))] = \int_{\mathbf{X}} h(\mathbf{X}) f(\mathbf{X})\, d\mathbf{X} \tag{4-6}$$

for any real-valued function $y(\mathbf{X})$. When the joint density $f(\mathbf{X})$ can be written as a simple product of marginal densities, it may then be possible to develop an expression for the transform of the distribution $f(y)$ as a simple function of the transforms of the component distributions $f_i(x)$. If the joint density is otherwise defined, the transform in Eq. (4-6) is still valid even though a simplification comparable to our previous product argument may not be possible.

The extension to discrete distributions and nonnegative random variables is straightforward. These will be left as exercises for the reader.

Linear Combinations

Consider an inspection station on an aircraft assembly line which uses a dye penetrant to check for cracks in nose fork castings. The cost of the dye used to inspect a casting is b dollars. The time required to perform the inspection is a random variable distributed by $f(t)$, where t is measured in minutes. The labor rate for the inspector at this station is a dollars per minute. The inspection cost for a single casting is then given by the random variable

$$c = at + b$$

The question now is, How can we use transforms to develop the distribution for inspection cost?

From Eq. (4-6) we can write

$$\mathscr{L}(g(c)) = E[e^{-s(at+b)}] = \int_t e^{-sat-sb} f(t)\, dt$$

The term e^{-sb} does not contain the random variable t and can thus be treated as a constant by pulling it outside the integral sign. Therefore

$$\mathscr{L}(g(c)) = e^{-sb} \int_t e^{-(sa)t} f(t)\, dt$$

The integral portion of this expression looks similar to the transform of the distribution of $f(t)$ except that the argument s has been replaced by the argument sa. It therefore follows that the transform for $g(c)$ can be written as a simple function of the transform of inspection time as

$$\mathscr{L}(g(c)) = e^{-sb} \mathscr{L}(f(t)) \Big|_{s=sa}$$

To consider a particular case, if the inspection time follows the exponential distribution

$$f(t) = ue^{-ut}$$

the transform for the distribution of cost is given by

$$\mathscr{L}(g(c)) = e^{-sb}\left(\frac{u}{s+u}\right)\bigg|_{s=sa} = \frac{ue^{-sb}}{sa+u} = \frac{(u/a)e^{-sb}}{s+(u/a)} \tag{4-7}$$

From the shifting theorem of Laplace transforms we know that

$$\mathscr{L}[f(t-k)U(t-k)] = e^{-ks}F(s)$$

where $F(s) = \mathscr{L}(f(t))$ and $U(t-k)$ is the unit step function beginning at $t = k$. We also know from our table of transform pairs that

$$\mathscr{L}(e^{mt}) = \frac{1}{s-m}$$

From these relationships it follows that the distribution of inspection cost must be

$$g(c) = \frac{u}{a}e^{-u(c-b)/a}, \qquad c \geq b \tag{4-8}$$

Now suppose that a production run of nose forks consists of n castings. The question now to be answered is, What is the distribution for the cost of producing a batch of nose forks? The random variable in this instance is

$$q = c_1 + c_2 + \cdots + c_n$$

where the c_i variables are identically distributed according to Eq. (4-8). Again employing the convolution property of transforms, it follows that

$$\mathscr{L}(h(q)) = [\mathscr{L}(g(c))]^n = \left[\frac{(u/a)e^{-sb}}{s+(u/a)}\right]^n \tag{4-9}$$

Inverting, one obtains from the transform of t^k and the complex shifting theorem

$$\mathscr{L}^{-1}\left[\left(\frac{(u/a)}{s+(u/a)}\right)^n\right] = \frac{(u/a)^n t^{n-1}e^{-(u/a)t}}{\Gamma(n)}$$

From the coefficient e^{-snb} we know that a shift on the time axis of nb units is involved. Finally, we have

$$h(q) = \frac{(u/a)^n(q-bn)^{n-1}e^{-(u/a)(q-bn)}}{\Gamma(n)}, \qquad q \geq bn \tag{4-10}$$

Equations (4-7)–(4-10), although developed in the context of a specific probability distribution, suggest a strategy for solving the general problem of finding distributions of linear combinations of independent random variables. Let

$$y = a_1 x_1 + a_2 x_2 + \cdots + a_n x_n + k \tag{4-11}$$

If each x_i is uniquely distributed, the transform for the distribution of $f(y)$ is given by

$$\mathscr{L}(g(y)) = e^{-sk}[\mathscr{L}(f_1(x))\mathscr{L}(f_2(x)) \cdots \mathscr{L}(f_n(x))]\Big|_{s_i = a_i s} \tag{4-12}$$

Here one temporarily identifies a unique s_i value for the transform of the ith distribution. After the indicated substitution takes place, the resulting function $\mathscr{L}(g(y))$ is a transform in the single variable s. For the special case of identically distributed values for x_i each multiplied by the same constant a, this expression reduces to

$$\mathscr{L}(g(y)) = e^{-sk}[\mathscr{L}(f(x))]^n\Big|_{s = as} \tag{4-13}$$

The inversion process will then yield appropriate distributions $g(y)$, which are defined as zero to the left of the point $y = k$.

Cumulative Distributions

In some applications it may be desirable to use the cumulative distribution or complementary cumulative distribution for a random variable rather than operate directly with the density function. The summation and integration properties of transforms are especially helpful in performing such calculations.

If x is a continuously distributed, nonnegative, random variable, the probability that x is less than some particular value of k is expressed as

$$F(x = k) = \Pr(x \le k) = \int_0^k f(x)\,dx \tag{4-14}$$

From the integration property of Laplace transforms we know that

$$\mathscr{L}\left(\int_0^x f(r)\,dr\right) = \frac{1}{s}\mathscr{L}(f(x)) \tag{4-15}$$

It therefore follows that the transform of the cumulative distribution can be obtained from the transform of the density by dividing by s. For example, suppose that

$$\mathscr{L}(f(x)) = \frac{a}{s + a}$$

It then follows that the transform of the cumulative distribution is

$$\mathscr{L}(F(x)) = \frac{a}{s(s + a)}$$

A similar relationship holds for integer-valued random variables when we employ geometric transforms. The transform of the cumulative distribution is given by

$$G(F(n)) = \sum_{n=0}^{\infty} \left(\sum_{i=0}^{n} f(i) \right) z^n = \frac{1}{1 - z} G(f(n)) \tag{4-16}$$

from the properties of the sum calculus. What we are suggesting is that the summation operation implied by the cumulative distribution be replaced by a simple multiplication by a function of z in the transform domain.

Random Sums of Discrete Random Variables

Recall that early in the last chapter we demonstrated that one could obtain the transform for the distribution of sums of independent random variables by multiplying the transforms of the distributions of the random variables contained in that sum. The problem we now want to examine concerns the case in which not only each term in the sum is a random variable, but also the number of terms in the sum is a random variable.

A problem of this sort arises in the context of inventory control systems, for example, where one may have data on the number of customers per day and the number of days required for delivery of an order (lead time). The problem is to develop an expression for the lead-time demand distribution. Let

$$y = x_1 + x_2 + \cdots + x_n \tag{4-17}$$

where the x_i values are identically distributed by $f(x)$ and n is distributed by $p(n)$. Furthermore, let us adopt the following notation for the geometric transforms of the distributions involved:

$$G_x(z) = G(f(x)) = \sum_{x=0}^{\infty} f(x) z^x \tag{4-18}$$

$$G_n(z) = G(p(n)) = \sum_{n=0}^{\infty} p(n) z^n \tag{4-19}$$

This notation emphasizes the fact that the transforms are functions of the variable z.

If we ignore for the moment the fact that n is a random variable, the sum in Eq. (4-17) is the same one we have examined several times before. The random variable y and its associated distribution are conditioned on n. If the value of n is fixed at, say, $n = 3$, we could write the conditional distribution for y as

$$h(y \mid n = 3) = f(x_1) * f(x_2) * f(x_3) = f(x)^{*3}$$

where the $(*)$ represents the convolution operation. The general case then follows as the n-fold convolution of $f(x)$:

$$h(y \mid n) = f(x)^{*n} \tag{4-20}$$

At this point it is helpful to once more invoke the convolution property of the geometric transform. We could think of the transform of $h(y \mid n)$ as a conditional transform. That is, it is the transform of the distribution for the random variable y conditioned upon the sum containing n terms. Let us write

$$G_y(z \mid n) = G(h(y \mid n)) = \sum_{y=0}^{\infty} h(y \mid n) z^y \tag{4-21}$$

But we know that the distribution of y for any fixed n is the n-fold convolution of $f(x)$, in which case the associated geometric transform can be written as

$$G_y(z \mid n) = [G_x(z)]^n \tag{4-22}$$

We can now return to the original problem, which was to find the marginal distribution of y. That is, find the distribution for y regardless of the value of n. From basic probability theory we know that

$$h(y) = \sum_n h(y, n)$$

where $h(y, n)$ is the joint distribution of y and n. But the joint distribution can be written as the product of a conditional and a marginal distribution as

$$h(y, n) = h(y \mid n) p(n)$$

It therefore follows that

$$h(y) = \sum_n h(y \mid n) p(n) \tag{4-23}$$

Using Eq. (4-23) to develop an expression for the transform of $h(y)$, we obtain

$$G_y(z) = \sum_y h(y) z^y = \sum_y \sum_n h(y \mid n) z^y p(n) \tag{4-24}$$

By interchanging the order of summation, we are led to

$$G_y(z) = \sum_n \left[\sum_y h(y \mid n) z^y\right] p(n) \tag{4-25}$$

The terms in the square brackets in Eq. (4-25) should look familiar. This is nothing more than the conditional transform defined in Eqs. (4-21) and (4-22). Using that information, we now have

$$G_y(z) = \sum_n G_y(z \mid n) p(n) = \sum_n [G_x(z)]^n p(n) \tag{4-26}$$

At this point we can take advantage of some mathematical insight to shorten the calculation implied by the summation of Eq. (4-26). We already know that the geometric transform of the distribution $p(n)$ is defined by

$$G_n(z) = \sum_n z^n p(n)$$

Comparing this expression with Eq. (4-26), a striking similarity is noted. The expression for $G_y(z)$ is identical in form to the expression for $G_n(z)$. The only difference is that the argument z in $G_n(z)$ has been replaced by the function $G_x(z)$. Let us note this relationship by writing

$$G_y(z) = G_n(z)\Big|_{z = G_x(z)} \tag{4-27}$$

where the vertical bar signifies that the argument z is to be replaced by the function $G_x(z)$. A slightly more compact expression using functional notation is

$$G_y(z) = G_n[G_x(z)] \tag{4-28}$$

We can summarize the above discussion by enumerating the following operational rules:

1. Develop an expression for the random sum of random variables as $y = x_1 + x_2 + \cdots + x_n$.
2. Find the transform for the distribution of the number of terms in the sum n. Designate this transform as $G_n(z)$.
3. Find the transform for the distributions of the independent, identically distributed random variables x contained in the sum. Designate this transform as $G_x(z)$.
4. Everyplace that the argument z appears in the transform $G_n(z)$, replace that argument with the entire transform $G_x(z)$.
5. The resulting function of z is the transform for the distribution of y. The distribution can be recovered by inverting the transform $G_y(z)$.

It might be well to note that the step by step procedure outlined above is predicated on the variables x_i being identically distributed. It is conceivable that one might be faced with a sum containing a random number of terms each one of which is uniquely distributed. In such an event, the simple substitution we used to obtain a closed form expression for the final transform is no longer meaningful. However, we can still employ the idea of a conditional transform which is weighted by the probability of n terms in the summation. That is,

$$G_y(z) = \sum_n G_y(z \mid n) p(n) \tag{4-29}$$

is valid even though the conditional transform $G_y(z \mid n)$ may no longer be a simple power of a single transform $G_x(z)$. In this situation

$$\begin{aligned} G_y(z \mid n = 1) &= G_{x_1}(z) \\ G_y(z \mid n = 2) &= G_{x_1}(z) G_{x_2}(z) \\ &\vdots \\ G_y(z \mid n) &= G_{x_1}(z) G_{x_2}(z) \cdots G_{x(n)}(z) \end{aligned} \tag{4-30}$$

and the existence of a neat closed form for $G_y(z)$ is no longer obvious.

Such a situation might arise in studying the route structure for a traveling salesman carrying a perishable sample. Suppose that he visits one city per day in the order 0, 1, 2, 3, 4. Sales per visit in city j are estimated to be distributed by $f_j(x)$. The distribution for the life of his sample goods in days is known to be $p(n)$. The problem is to estimate the distribution of total sales from his current sample and planned itinerary. Assume the following data:

$$p(n) = \begin{cases} 0.1, & n = 1 \\ 0.2, & n = 2 \\ 0.3, & n = 3 \\ 0.4, & n = 4 \end{cases}$$

$$f_j(x) = \frac{(10 - j)^x e^{-(10-j)}}{x!}, \qquad j = 0, 1, 2, 3, 4$$

The geometric transform for the distribution of sales in city j is given by

$$G(f_j(x)) = e^{-(10-j)(1-z)}$$

The transform for the distribution of total sales for the trip is given by

$$\begin{aligned} G_y(z) &= \sum_{n=1}^{4} \left[\prod_{k=1}^{n} G(f_{k-1}(x)) \right] p(n) \\ &= 0.1e^{-10(1-z)} + 0.2e^{-19(1-z)} + 0.3e^{-27(1-z)} + 0.4e^{-34(1-z)} \end{aligned}$$

Example Data taken from a particular inventory system are not sufficient to provide adequate statistical estimates for the necessary probability distributions. However, they do indicate that the number of days required to receive a replenishment order (lead time) never exceeds 10. Furthermore, the average lead time is 4 days. Lacking other data, the analyst might conclude that the binomial distribution is a reasonable model. A study of the number of customers indicates an average of 20 customers per day. Although the exact form of the distribution is unknown, the analyst may feel that this is a completely random process. Furthermore, comparing the existing customer population with those reported in the inventory literature could lead him to hypothesize that the Poisson distribution with a mean of 20 may be an adequate model for daily customer traffic. In order to protect the system against shortages, it is necessary to know the distribution for the number of customers who will require service during the time it takes to receive a replenishment shipment.

Let y be the number of customers in a lead time, x_i the number of customers on day i, n the number of days in a lead time,

$$p(n) = \frac{10!}{n!\,(10-n!)}(0.4)^n(0.6)^{10-n}, \qquad 0 \le n \le 10$$

and

$$f(x) = \frac{(20)^x e^{-20}}{x!}, \qquad x \ge 0$$

We are now in a position to apply the operational procedures noted in the last section.

1. The expression for lead-time demand is

$$y = x_1 + x_2 + \cdots + x_n$$

2. The transform for $p(n)$ is

$$G_n(z) = \sum_{n=0}^{10} \frac{10!}{n!\,(10-n!)}(0.4z)^n(0.6)^{10-n} = (0.4z + 0.6)^{10}$$

3. The transform for $f(x)$ is

$$G_x(z) = \sum_{x=0}^{\infty} \frac{(20z)^x e^{-20}}{x!} = e^{-20(1-z)}$$

4. Apply Eq. (4-28) to obtain

$$\begin{aligned} G_y(z) &= G_n[G_x(z)] \\ &= (0.4z + 0.6)^{10}\Big|_{z=\exp(-20(1-z))} \\ &= [0.4\exp(-20(1-z)) + 0.6]^{10} \end{aligned}$$

5. The distribution for y can now be recovered by inverting the transform if possible. In the event that only the moments of the distribution are needed, they can be recovered by taking appropriate derivatives of the geometric transform $G_y(z)$ evaluated at $z = 1$ as outlined in the previous chapter.

Higher Order Nesting

One might think of the transform for the distribution of a random sum of random variables as being one in which the transform for the distribution of components of the sum is "nested" within the transform for the distribution of the number of terms in that sum. The obvious next question concerns whether or not that nesting process can be continued to develop closed form expressions for transforms involving more complex random variables expressed as a function, of a function, of a function, ad nauseam.

To illustrate the point, suppose that each customer in our previous inventory example requires a random number of items k distributed by $f(k)$. If we now consider d_i to be the number of items demanded in day i, we can write

$$d_i = k_1 + k_2 + \cdots + k_x$$

where x represents the number of customers per day. Conceptually, we view the system as one in which the first customer of the day demands k_1 items, the second k_2, and so forth, through the x customers served on day i. The resulting expression is a random sum of random variables. From our previous work we know that the transform for the distribution of d_i can be expressed as

$$G_d(z) = G_x[G_k(z)]$$

The number of items w demanded in a lead time can then be viewed as the number demanded on day one plus the number on day two and so forth up to the total number of days in the lead time. Symbolically, we have

$$w = d_1 + d_2 + \cdots + d_n$$

where n represents the number of days in the lead time. Here again we have a simple random sum with the associated transform

$$G_w(z) = G_n[G_d(z)]$$

But we previously developed an expression for $G_d(z)$ which was itself a nested transform. It therefore follows that we could write an expression for the transform of the distribution for items demanded in a lead time as

$$G_w(z) = G_n[G_x[G_k(z)]] \tag{4-31}$$

To illustrate the technique, suppose that we have data to indicate that the number of items demanded per customer can be described by the geometric distribution

$$f(k) = (1 - p)p^k, \qquad k = 0, 1, \ldots, \infty$$

The transform of this distribution is

$$G_k(z) = \sum_{k=0}^{\infty} (1 - p)(pz)^k = \frac{1 - p}{1 - pz}$$

Using the same distributions as before for x and n, we have

$$G_x(z) = \exp(-20(1 - z)) \qquad \text{and} \qquad G_n(z) = (0.4z + 0.6)^{10}$$

The nesting argument of Eq. (4-31) now suggests that we replace the transform variable z in the expression for $G_n(z)$ with an entire transform $G_x(z)$. This leads us to

$$G_w(z) = (0.4z + 0.6)^{10} \Big|_{z = \exp(-20(1-z))|_{z=(1-p)/(1-pz)}}$$

$$= \left[0.4 \exp\left(-20\left(1 - \frac{1-p}{1-pz}\right)\right) + 0.6\right]^{10} \tag{4-32}$$

Although complete inversion of an expression such as Eq. (4-32) may be exceedingly difficult, the reader should bear in mind that one often needs only the lower order moments or perhaps just the first few terms of the probability vector. In spite of its formidable appearance, the above expression is a simple geometric transform in the argument z. As such, it can be differentiated to recover as many moments or as many terms of the probability distribution as we have patience for taking derivatives. As we noted in the last chapter,

$$E(w) = \frac{dG_w(z)}{dz}\Big|_{z=1}$$

$$\text{var}(w) = \frac{d^2G_w(z)}{dz^2}\Big|_{z=1} + \frac{dG_w(z)}{dz}\Big|_{z=1} - \left[\frac{dG_w(z)}{dz}\Big|_{z=1}\right]^2$$

and

$$f(w = j) = \frac{1}{j!}\frac{d^jG_w(z)}{dz^j}\Big|_{z=0}$$

It is now a simple matter to extend this approach to as many levels of nesting as seems desirable in any given problem context. For example, we might want to know the amount of income generated during a lead time in our inventory example. Here let us suppose that the income b on any given item is distributed by $f(b)$. It then follows that income c per customer is given by

$$c = b_1 + b_2 + \cdots + b_k$$

income per day d is

$$d = c_1 + c_2 + \cdots + c_x$$

and income per lead time t is

$$t = d_1 + d_2 + \cdots + d_n$$

Applying the nesting arguments, it follows that

$$G_t(z) = G_n[G_d(z)] = G_n[G_x[G_c(z)]] = G_n[G_x[G_k[G_b(z)]]] \qquad (4\text{-}33)$$

which is once more a function of the single transform variable z.

It should be obvious to the reader that any attempt to memorize the particular order in which the transforms are to be nested for higher order nesting problems could quickly become an exercise in futility. The proper approach, as illustrated in our inventory example, is to break the problem into a series of first-order random sums, defining new random variables at each level as required. The nesting is then easily performed one level at a time until the desired compound expression is obtained.

As an aid for tracing the interrelationships among the pairs of random variables constituting the first-order random sums of random variables, let us construct a logic network with the following characteristics.

1. Every node represents a random variable.
2. Each intermediate node has two branches emanating from it to indicate that the variable at that node is a function of two random variables, the number of terms in the sum and the value of individual terms.

(i) The right branch leads to a node representing the number of terms in the summation.

(ii) The left branch leads to a node representing a single term in the summation.

3. Only the transforms associated with those random variables located at the initial node and at terminal nodes will appear in the nested transform.
4. The transform for the compound random variable is constructed functionwise by nesting transforms in order, beginning with the rightmost terminal node at each level in the network. Nesting is exhausted at each level by moving from right to left. Levels are defined by those variables on the extreme left edge of the network. The transform is complete when all levels have been traversed.

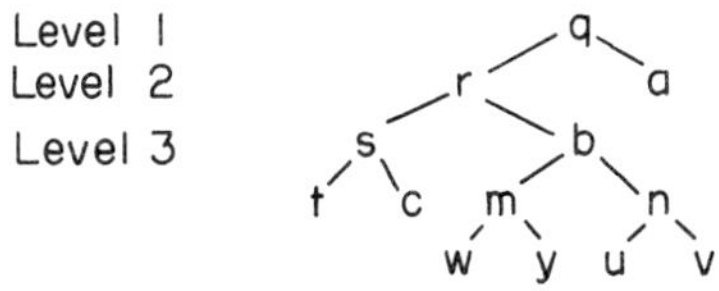

Figure 4-1 A compound random variable.

Let us illustrate the technique by means of the network in Figure 4-1. This network was constructed from the following set of equations:

$$q = f(r, a) = r_1 + r_2 + \cdots + r_a$$

$$r = f(s, b) = s_1 + s_2 + \cdots + s_b$$

$$b = f(m, n) = m_1 + m_2 + \cdots + m_n$$

$$n = f(u, v) = u_1 + u_2 + \cdots + u_v$$

$$m = f(w, y) = w_1 + w_2 + \cdots + w_y$$

$$s = f(t, c) = t_1 + t_2 + \cdots + t_c$$

Beginning with level one, the right variable is a, which happens also to be a terminal node. This indicates that the outermost transform will be $G_a(z)$. (We will drop the z-argument temporarily to facilitate the notation.) The transform then starts as

$$G_q(z) = G_a[\cdots]$$

where the terms in the bracket are yet to be determined.

The right branch at level two leads to further intermediate branches. Following that path, the rightmost terminal pair to the left on those branches emanating from the second level gives two more transforms for the nest in proper order:

$$G_q(z) = G_a[G_v[G_u[G_y[G_w[\cdots]$$

Having exhausted all secondary branches from this level, we now move to level three, which branches only to the terminal nodes c and t. This then completes our expression as

$$G_q(z) = G_a[G_v[G_u[G_y[G_w[G_c[G_t(z)]]]]]]$$

Continuous Variables in Random Sums

In our previous discussions of random sums all of the random variables involved were assumed to be integer-valued. The geometric transform was used to characterize their accompanying discrete probability distributions.

This led us via some mathematical trickery to obtain closed form expressions for certain infinite sums by nesting geometric transforms. The question now to be explored is what recourses are available to us if certain of the variables in a random sum are continuous rather than discrete.

Consider the yield to be obtained from a certain chemical process. The number of gallons of product produced in a single day, x, is a continuously distributed random variable. The life of the process depends upon the durability of a catalyst used in the operation. The catalyst remains useful for a number of days n, where, as a first approximation, n is assumed to be discrete. This is the simplest case of a sum of continuously distributed, independent random variables in which the number of terms remains discrete. That is,

$$y = x_1 + x_2 + \cdots + x_n \tag{4-34}$$

where x_i is distributed by $f_i(x_i)$, a continuous distribution. Further, suppose that all variables are nonnegative.

The transforms to be employed for purposes of illustration will be the Laplace transform for the $f_i(x_i)$ distributions and the geometric transform for the $p(n)$ distribution:

$$\mathscr{L}[f_i(x_i)] = F_{x_i}(s) = \int_0^\infty e^{-sx} f(x)\,dx \tag{4-35}$$

and

$$G(p(n)) = G_n(z) = \sum_{n=0}^{\infty} z^n p(n) \tag{4-36}$$

Note that the transform argument for $G(p(n))$ is not the same variable as the transform argument for $\mathscr{L}[f_i(x_i)]$. The symbols z and s are used to uniquely identify the type of transform existing at any stage of the calculation. This is different from our last efforts, in which all transforms were geometric and could be characterized by the single complex variable z.

If one views Eq. (4-34) for any particular value of n, the problem reduces to a simple sum of continuously distributed random variables. From our previous work we know that the transform for the distribution of such a sum can be obtained from the product of the transforms of distributions for variables contained in the sum. Again using the concept of a conditional transform, let us write the Laplace transform for y given a particular value of n as

$$\mathscr{L}[h(y \mid n)] = F_y(s \mid n) = [F_x(s)]^n \tag{4-37}$$

Since the number of terms n is also a random variable, we can obtain the unconditional transform by summing the weighted Laplace transforms. That is,

$$\mathscr{L}[h(y)] = \sum_{n=0}^{\infty} F_y(s \mid n)p(n) = \sum_{n=0}^{\infty} [F_x(s)]^n p(n) \tag{4-38}$$

Up to this point the argument is completely analogous to that used for the discrete variables in Eq. (4-26). The only difference is that we have a sum of weighted Laplace transforms rather than a sum of weighted geometric transforms. This obviously implies that the resulting transform for $h(y)$ is a Laplace transform.

Now let us once more employ our mathematical insight to note the close resemblance of Eq. (4-38) to the formal definition of the geometric transform of the distribution $p(n)$. If one replaces the argument z in the definition shown in Eq. (4-36) with the Laplace transform of $f(x)$, the expression for $\mathscr{L}(h(y))$ results. Therefore the transform for the distribution of total process yield can be written as

$$\mathscr{L}[h(y)] = G[p(n)]\Big|_{z=\mathscr{L}[f(x)]} = G_n[F_x(s)] \tag{4-39}$$

The reader should take particular note of Eq. (4-39). Even though the first transform in the nest is a geometric transform, it is being employed only as a means of obtaining a closed form expression for $\mathscr{L}(h(y))$. The resulting transform is a Laplace transform in the argument s. It is not a geometric transform at all. The transform variable z does not appear in the end expression. A moment's reflection should offer some intuitive support for this result. We have added continuously distributed random variables and should therefore expect the sum of these to also be continuous. The final transform must be Laplace.

As long as the number of terms in any summation remains a discrete random variable, the inclusion of a continuously distributed random variable in higher order nesting problems poses no new difficulties. The nesting sequence proceeds exactly as it did for the purely discrete case as far as determining the proper order of function arguments goes. The only thing to remember is that a single continuous variable in the problem will make the resulting random sum continuous. The final transform will be a Laplace transform. The seeming presence of geometric transforms occurs only because we have employed them as mathematical identities to find closed form expressions for weighted Laplace transforms.

Example Consider a plant maintenance department which must service the automatic transfer machines on a production line. The number of service calls per day can be described by a Poisson distribution with a mean of 4. The time required to make a repair follows the negative exponential distribution with a mean of $\frac{1}{2}$ hour. The problem is to find a distribution for the number of man-hours of repair time used per day.

Let

$$d = t_1 + t_2 + \cdots + t_n$$

where

$$f(t) = 2e^{-2t}, \quad t \geq 0, \qquad \text{and} \qquad p(n) = \frac{4^n e^{-4}}{n!}, \quad n = 0, 1, 2, \ldots$$

The associated transforms are

$$\mathscr{L}(f(t)) = \int_0^\infty e^{-st} 2e^{-2t}\, dt = \frac{2}{s+2} \qquad \text{and} \qquad G(p(n)) = e^{-4(1-z)}$$

From Eq. (4-39) it follows that

$$\mathscr{L}(h(d)) = G_n[F_t(s)] = \exp\left[-4\left(1 - \frac{2}{s+2}\right)\right] = \exp -\frac{4s}{s+2}$$

Since we now have a Laplace transform, the moments of $h(d)$ can be recovered by taking appropriate derivatives with the argument s set equal to zero. That is,

$$E(d^k) = (-1)^k \left.\frac{d^k \mathscr{L}(h(d))}{ds^k}\right|_{s=0} \tag{4-40}$$

The density $h(d)$ can be recovered by inverting the Laplace transform.

Integration for Random Processes

In the previous section we introduced techniques for handling random sums when the components in the sums are continuously distributed random variables. We did demand, however, that the number of terms in the summation be an integer-valued random variable. What we now want to explore is the continuous analog of the random sum. One could think of this as an integration process. The analogy involves adding an infinite number of random variables each taken over a small period Δt as Δt approaches zero. This is a limiting case for the random sum in which each variable was defined over a period of unity. What we now seek is a distribution for the limiting version of the summation or the integration of a random process.

Such a situation might arise in the evaluation of tooling alternatives for aluminum fairings used in aircraft production. Suppose that management is considering the use of forming dies made from tooling Masonite as one alternative. Because of the high manual labor content, production rate is highly variable and can best be described in terms of a probability distribution. Furthermore, the life of the Masonite forming dies depends upon the quality of the Masonite, existing temperature and humidity conditions, and the care used by operating personnel. The net effect is that the total production time obtainable from a set of dies is a continuously distributed random variable.

What is now required is an estimate of the probability distribution for the total number of fairings produced from a single set of dies. To express this problem in symbols, consider the case in which the distribution for the number of units x produced by a set of dies is a function of the duration t of the life of the dies, where t is itself a continuous random variable. Suppose that x is discrete and let the conditional distribution for x be given by $p(x \mid t)$. The joint distribution for the continuous variable t and the discrete variable x is then given by

$$h(x, t) = p(x \mid t) f(t) \tag{4-41}$$

In order to find the marginal distribution for x, it is necessary to integrate over t:

$$h(x) = \int_0^\infty h(x, t)\, dt = \int_0^\infty p(x \mid t) f(t)\, dt \tag{4-42}$$

Since x is discrete, we can define the geometric transform for $h(x)$ as

$$\begin{aligned} G(h(x)) = \sum_{x=0}^{\infty} h(x) z^x &= \sum_{x=0}^{\infty} \left[\int_0^\infty p(x \mid t) f(t)\, dt \right] z^x \\ &= \int_0^\infty \left[\sum_{x=0}^{\infty} p(x \mid t) z^x \right] f(t)\, dt \end{aligned} \tag{4-43}$$

But the terms in the brackets in Eq. (4-43) are nothing more than the geometric transform of the conditional distribution. It therefore follows that

$$G(h(x)) = \int_0^\infty G_x(z \mid t) f(t)\, dt \tag{4-44}$$

Equation (4-44) is the continuous analog of Eq. (4-23). We are now integrating a weighted transform conditioned on a continuous variable t, where before we were summing a weighted transform conditioned on a discrete variable n. As before it is sometimes possible to shorten the calculation by recognizing similarities to transforms with different arguments.

This calculation is perhaps best illustrated by the Poisson process. Suppose that the arrival of ships for passage through the locks of a canal is governed by a Poisson process with a mean rate of u ships per hour. Further suppose that the time required to effect repairs to the locks should a failure occur is exponentially distributed with a mean of h hours. If the locks fail when there are no ships in the queue, the question is, How many ships will be present when the repairs are completed?

Let

$$p(x \mid t) = \frac{(ut)^x e^{-ut}}{x!} \qquad \text{and} \qquad f(t) = \frac{1}{h} e^{-t/h}$$

Then from Table 3-2

$$G_x(z \mid t) = e^{-ut(1-z)} \qquad \text{and} \qquad \mathscr{L}[f(t)] = \frac{1}{hs + 1}$$

Applying Eq. (4-44), it follows that

$$G(h(x)) = \int_0^\infty e^{-ut(1-z)} \frac{1}{h} e^{-t/h} \, dt \tag{4-45}$$

Note the close resemblance of this equation to the definition of the Laplace transform for $f(t)$. If we replace the terms $u(1 - z)$ in the first expression by the variable s, Eq. (4-45) becomes the Laplace transform of $f(t)$. That is,

$$G(h(x)) = \int_0^\infty e^{-st} \frac{1}{h} e^{-t/h} \, dt = \mathscr{L}(f(t)) \Big|_{s=u(1-z)} \tag{4-46}$$

The complete expression for the distribution of the number of ships delayed during repair is then

$$G(h(x)) = F_t(s) \Big|_{s=u(1-z)} = F_t[u(1 - z)] = \frac{1}{hu(1 - z) + 1} \tag{4-47}$$

Note that the transform $G(h(x))$ is a geometric transform, not a Laplace transform. We have only used our knowledge of the form of the integral of the Laplace transform to avoid integrating a new function. This is similar to the reasoning we previously used when we developed a Laplace transform expressed as the geometric transform of another function with its argument replaced by still another Laplace transform. We should also hasten to point out that the substitution in Eq. (4-46) was made possible only because of the special form of the geometric transform of the Poisson distribution. It is not generally true that the conditional transform will be expressed as a power of e. In the general case we must be content to execute the integration implied by Eq. (4-44).

Permitting x to be a continuous random variable poses no new conceptual difficulties. The only change is that the summation of Eq. (4-43) is replaced by an integration argument. That is,

$$\mathscr{L}(h(x)) = \int_0^\infty h(x)e^{-sx}\,dx = \int_0^\infty \left[\int_0^\infty p(x\mid t)f(t)\,dt\right]e^{-sx}\,dx$$

By changing the order of the integration, we obtain

$$\mathscr{L}(h(x)) = \int_0^\infty \left[\int_0^\infty p(x\mid t)e^{-sx}\,dx\right]f(t)\,dt \tag{4-48}$$

But the terms in brackets define the Laplace transform for the conditional distribution $p(x \mid t)$. It then follows that $\mathscr{L}(h(x))$ is the integral of a continuously weighted conditional Laplace transform:

$$\mathscr{L}(h(x)) = \int_0^\infty \mathscr{L}(p(x\mid t))f(t)\,dt = \int_0^\infty F_x(s\mid t)f(t)\,dt \tag{4-49}$$

Moments for Compound Random Sums

So far the nesting arguments introduced for random sums of random variables have been seen to be conceptually simple techniques which may lead to impossibly complicated expressions for the transform of the distributions we may be seeking. The main advantage of the technique as discussed is to facilitate the data gathering process. For example, in the inventory context it may be a relatively simple matter to collect data on days in a lead time and as a separate effort to collect data on the number of customers per day. If distributions can be fit to these data separately, we now in theory have a method to find the compound distribution. We can solve the problem by nesting transforms even though it might be virtually impossible to assemble adequate data from the accounting system to empirically establish the lead-time demand distribution directly.

As a matter of operational reality, however, one must soon face up to the fact that data systems are rarely complete enough to permit adequate estimates of the component probability distributions. The analyst must often be satisfied with an estimate of the mean and possibly the variance of the component distributions. We will now turn our attention to ways of using this limited information to find the moments of the compound distributions of interest. As an auxiliary benefit, the resulting equations can be used even when the entire component distributions may be known in lieu of the often tedious task of differentiation. What we are suggesting is that the differentiation be done once for the general case rather than repeatedly as each unique problem arises.

In the context of determining demand levels for inventory control models, one may be able to extract enough data from customer orders to estimate the mean and variance for the number of units demanded by a customer. A brief work sampling study may be sufficient to estimate the number of customer inquiries, including both those served and those denied service, in a day's time. A study of purchase orders and receiving documents may yield an estimate for the mean and variance of the time required to receive a replenishment order. The problem is to use these meager statistics to develop a meaningful estimate for the number of units demanded while the company awaits a replenishment order. In many such cases the data are insufficient to warrant fitting known probability distributions to each of the variables, let alone finding the distribution for the composite.

Let us begin with the first-order nesting for a random sum of discrete random variables as represented by Eq. (4-28). That is for n terms in a sum of x variables

$$G(h(y)) = G_y(z) = G_n[G_x(z)]$$

From our previous work we know that the expected value of y can be obtained from the first derivative of $G(h(y))$ evaluated at $z = 1$. Let us now take that derivative by using the chain rule for the derivative of implicit functions:

$$\frac{dG_y(z)}{dz} = \left(\frac{\partial G_n}{\partial G_x}\right)\left(\frac{\partial G_x}{\partial z}\right)$$

or

$$G_y'(z) = G_n'[G_x(z)]G_x'(z) \tag{4-50}$$

To evaluate this expression at $z = 1$, we will use the following properties of geometric transforms:

$$G_k(z)\Big|_{z=1} = 1 \qquad \text{and} \qquad G_k'(z)\Big|_{z=1} = E(k)$$

It then follows that

$$G_y'(z)\Big|_{z=1} = G_n'(1)G_x'(1)$$

or

$$E(y) = E(n)E(x) = \mu_n \mu_x \tag{4-51}$$

This says that if we need to find the expected value for a random sum of discrete random variables, we need only know the expected value of the number of terms and the expected value of individual terms in the sum. This product gives the desired answer.

In many practical problems it may also be desirable to find the variance for a random sum. To perform that calculation using the geometric transforms, we recall that

$$\text{var}\,(y) = [G_y''(z) + G_y'(z) - [G'(z)]^2]\Big|_{z=1} \tag{4-52}$$

Beginning with the first derivative we have

$$\begin{aligned} G_y''(z) &= \frac{d}{dz}[G_y'(z)] = \frac{d}{dz}[G_n'[G_x(z)]G_x'(z)] \\ &= G_n''[G_x(z)][G_x'(z)]^2 + G_x''(z)G_n'[G_x(z)] \end{aligned} \tag{4-53}$$

To evaluate this expression at $z = 1$, it is helpful as an intermediate step to reorder the computational equation for variance as a function of derivatives of the transform. This will permit us to replace the second derivative terms with functions of means and variances of the distributions involved, namely

$$G_k''(1) = \text{var}\,(k) + [E(k)]^2 - E(k) \tag{4-54}$$

It then follows from Eqs. (4-52) and (4-53) that

$$\begin{aligned} \text{var}\,(y) &= G_n''(1)[E(x)]^2 + G_x''(1)E(n) + E(n)E(x) - [E(n)E(x)]^2 \\ &= [\sigma_n^2 + \mu_n^2 - \mu_n]\mu_x^2 + [\sigma_x^2 + \mu_x^2 - \mu_x]\mu_n + \mu_n\mu_x - [\mu_n\mu_x]^2 \end{aligned}$$

or

$$\sigma_y^2 = \mu_x^2\sigma_n^2 + \mu_n\sigma_x^2 \tag{4-55}$$

These results now permit us to calculate the mean and variance for random sums of random variables as simple functions of the means and variances of the component distributions. This result is extremely helpful in analyzing systems in which the first two moments may be the most information we have about the probability distributions. It is also useful in those contexts in which the complete distributions are known since it avoids the necessity of taking sometimes awkward derivatives.

The extension to continuous functions is straightforward. Equations (4-51) and (4-55) still hold even when x is a continuous variable. The key to proving that statement is to recognize after applying the chain rule that

$$\mathscr{L}(f(t))\Big|_{s=0} = 1$$

The details of the proof are left as an exercise for the reader.

The extension of these results to higher order nesting problems proceeds in straightforward fashion. For example, if the transform of interest is

$$G_y(z) = G_a[G_b[G_c(z)]] \tag{4-56}$$

we simply employ the associative law of transforms and solve the problem in two stages. Define

$$G_d(z) = G_b[G_c(z)] \tag{4-57}$$

Then rewrite $G_y(z)$ as

$$G_y(z) = G_a[G_d(z)] \tag{4-58}$$

From our results on the first-order nesting in Eq. (4-57) we know that

$$\mu_d = \mu_b \mu_c \qquad \text{and} \qquad \sigma_d^2 = \mu_c^2 \sigma_b^2 + \mu_b \sigma_c^2$$

Furthermore, using the same results for Eq. (4-58) and then making the above substitutions for the intermediate variable d, we obtain

$$\mu_y = \mu_a \mu_d = \mu_a \mu_b \mu_c \tag{4-59}$$

and

$$\sigma_y^2 = \mu_d^2 \sigma_a^2 + \mu_a \sigma_d^2 = (\mu_b \mu_c)^2 \sigma_a^2 + \mu_a(\mu_c^2 \sigma_b^2 + \mu_b \sigma_c^2) \tag{4-60}$$

One can extend these results to even higher order nesting problems with repeated applications of Eqs. (4-51) and (4-55). The algebra becomes more tedious but the concept is straightforward. The end result provides expressions for the mean and variance of any multilevel random sum as functions of the means and variances of the component distributions.

Example A marketing representative is attempting to estimate the mean and variance for the amount of beer consumed by residents of a "wet dorm" during a school term. Data from a brewery survey on student drinking habits indicate that a student who drinks will consume an average of 1.4 pints of beer per day. The variance for individual consumption is 0.09. The university newspaper has compiled data which indicate that a wet dorm will contain an average of 750 students who drink. The range in the newspaper data is 600–900 students. Lacking other data, let us hypothesize that the range of 300 represents six sigma units. Because of periodic campus unrest the university police feel that the length of a school term these days is a random variable. Their estimate is that the average term will be 65 days with a variance of 25. The problem is to estimate consumption for the school term:

Let c be the amount consumed by a single student, b the number of drinking students in a wet dorm, a the number of days in a school term, m the amount consumed by a dorm in 1 day, and n the amount consumed by a dorm in a term. Then

$$m = c_1 + c_2 + \cdots + c_b \qquad \text{and} \qquad n = m_1 + m_2 + \cdots + m_a$$

from which

$$G_n(z) = G_a[G_b[G_c(z)]]$$

Applying the results of Eqs. (4-59) and (4-60), we obtain

$$\begin{aligned} E(n) &= \mu_a \mu_b \mu_c = (65)(750)(1.4) = 68{,}250 \text{ pints} \\ \sigma_n^2 &= (\mu_b \mu_c)^2 \sigma_a^2 + \mu_a(\mu_c^2 \sigma_b^2 + \mu_b \sigma_c^2) \\ &= [(750)(1.4)]^2(25) + 65[(1.4)^2(2500) + 750(0.09)] \\ &= 27.88(10^6) \end{aligned}$$

If we are now bold enough to make the big assumption that all the world is normal, we can use these moments to make any probability estimates that might be desired.

Problems

1. Develop an equation similar to (4-6) for the case of integer-valued random variables, using the geometric transform.

2. Prove that the moments for a function of a random variable can be obtained by differentiating the transform of Eq. (4-6).

3. A carnival operator has a new game of chance to put on the midway this season. Each player must put up b dollars to enter the game. This entitles him to n shots at a target. Each time the target is hit, the customer is entitled to purchase one item of his choice from the merchandise on display. For each item selected, the player is charged an additional a dollars. The guns are rigged such that the probability of a hit on any one shot is 0.8.

(a) Find an appropriate transform for the distribution of dollars received per customer.

(b) If the number of customers expected in a day's time has a mean of c and a variance of v, find the mean and variance for total daily income.

4. An ocean cruise ship line is developing specifications for their new flagship. The shipbuilder can provide a total number of C corridors with cabins. Each corridor of cabins will be designated as first-class or tourist. A first-class corridor can accommodate six cabins. A tourist corridor can accommodate ten cabins. From past data the line has developed a discrete probability function $f_1(x)$ for the frequency with which x customers desire first-class passage on each cruise. They estimate p_1 as the net profit from selling a first-class passage. Similarly, $f_2(y)$ is the frequency with which y customers desire tourist passage and p_2 is the net profit from each tourist.

(a) Suppose that the distributions f_1 and f_2 are for cabin demand where each of the random variables x and y represents a group of customers. The size of a first-class group is a random variable distributed by $g_1(q)$. The size of a tourist group is distributed by $g_2(r)$. Develop an expression for the probability distribution, expressed as a function of the f and g distributions, for total number of passengers seeking passage on a cruise.

(b) Suppose that each cabin accommodates only one customer. Write the appropriate expression for expected profit and explain how you would find an optimal cabin allocation. State all assumptions.

5. A sponsor for fund raising dinners has developed a novel forecasting scheme whereby he looks for patriotic lapel pins worn by attendees. He feels that these pins are indicators of party affiliation. His data indicate that eight out of ten pin-wearers will be members of the speaker's party and the probability that any one attendee wears a pin is 0.3. Relying on past experience, he feels safe in stating that in 95% of the campaigns the average number of attendees at a fund raising dinner will fall between 400 and 792. These data were developed by observing attendance at 100 dinners. Further study of these data reveals that a pin wearer can be expected to contribute an average of $20 to the party with a variance of 16.

(a) Find the mean and variance for number of pin wearers observed at a fund raising dinner attended by 200 persons.

(b) Find a transform for the number of party members at a dinner attended by 200 persons.

(c) Find the mean and variance for number of party members at a dinner attended by 200 persons.

(d) Find the mean and variance for the dollar amount of contributions collected at a random dinner, assuming that only pin-wearers contribute.

(e) The sponsor feels that the number of attendees at tonight's special invitational dinner will follow a Poisson distribution with mean 10 and that each attendee will drink an amount distributed exponentially with a mean of 0.25 gallon.

(i) Develop an appropriate transform for the distribution of amount of drink consumed at the dinner.

(ii) Use the resulting transform to find the mean and variance of the consumption distribution.

6. The number of gallons of gasoline delivered to a service station customer is believed to be continuously distributed by the uniform pdf in the interval of 4–20 gallons. The number of customers per day follows a Poisson distribution with standard deviation of 5. The number of days between deliveries from the bulk fuel depot is distributed by

$$f(n) = \frac{5!}{(n-5)!\,(10-n)!}(0.3)^{n-5}(0.7)^{10-n}, \qquad n = 5, 6, \ldots, 10$$

(a) Develop an appropriate transform in closed form for each of the distributions given above.

(b) Find an appropriate transform for the number of gallons of gas sold between deliveries. What type of transform is it?

(c) Find the transform of the probability that there will be more than x gallons sold between bulk delivery periods.

7. Prove that Eqs. (4-51) and (4-55) are valid when x is a continuously distributed random variable.

8. The vice squad keeps statistics on the number of gambling arrests made on any given day. Their records show that on Saturday night the mean and variance for number arrested are μ_a and σ_a^2, respectively. When a gambler is arrested, he is either held in jail or released on bond. A typical group of n arrests will find rn released on bond. The duration of the jail stay for those held has a mean of μ_j days and a variance of σ_j^2. Each day a gambler is held in jail reduces his income by an amount y, where y has a mean of μ_y and variance σ_y^2. Find the mean and variance for the loss of income to the local gambling trade as a result of one Saturday night's vice squad activities. Assume all random variables are discrete.

9. Consider the problem of a security firm attempting to provide protection against truck hijackers. It has been estimated that that time between transfers of single crates from one vehicle to another will exceed t minutes with probability e^{-10t}. Furthermore, a crate will contain n items with probability

$$p(n) = \begin{cases} 0.5, & n = 100 \\ 0.3, & n = 120 \\ 0.2, & n = 150 \end{cases}$$

The length of time in minutes that a hijack operation will go undetected is gamma-distributed with parameters $a = 14$ and $b = 3$. It is necessary to identify the probability distribution for number of items obtained per hijack, prior to detection.

(a) Develop an appropriate transform for the items-per-hijack distribution. Include all functions but leave in integral form.

(b) Identify the transform obtained in part (a), i.e., is it geometric, Laplace, Fourier, or what?

(c) Find the mean and variance for number of items obtained per hijack.

10. Consider the problem of a customs agent whose job it is to confiscate a low grade of ethyl alcohol being smuggled across the border by rail. The alcohol is shipped in drums stashed carefully among similar drums in carloads of paint thinner. Due to the volatile nature of the alcohol, the agent estimates that the number of gallons per drum will be distributed by

$$f_1(a) = 0.05, \qquad 30 \le a \le 50$$

From a study of smuggling patterns, he has established that the number of drums per carload is likely to be distributed by

$$f_2(d) = 0.9(0.1)^d, \qquad d = 0, 1, 2, \ldots$$

Thinner shipping waybills show that the number of carloads of thinner in a train is distributed by

$$f_3(c) = \frac{8^{c+2}}{64\Gamma(c+1)e^8}, \qquad c = 0, 1, 2, \ldots$$

Since the agent moves around at random, the length of time he may spend at any border railway station is gamma-distributed $[f_4(t)]$ with a mean of 24 hours and a variance of 8. Due to the random nature of freight train schedules described by $f_5(t)$, he estimates the probability that time between arrivals of trains at a station will be less than t hours to be $1 - e^{-0.5t}$. All confiscated alcohol must be stored at a cost of \$4/gallon in a customs warehouse until the agent completes his tour of raids. The number of raids on a tour is distributed by

$$f_6(n) = \begin{cases} 0.5, & n = 2 \\ 0.3, & n = 4 \\ 0.2, & n = 6 \end{cases}$$

(a) Write an appropriate transform in closed form for each of the distributions $f_1(n), f_2(n), \ldots, f_6(n)$.

(b) Find an expression, using symbols only, for the transform of the distribution of number of gallons confiscated per raid.

(c) Find the transform for the distribution of number of gallons per train. What type of transform is it?

(d) Find a closed form expression for the transform of the distribution of the number of gallons of alcohol to be stored in the warehouse. What type of transform is it?

(e) Find the expected value of cost of storing alcohol confiscated on a tour.

11. The concession operator for a carnival ride insists that the equipment be fully loaded before he will start the ride. The capacity of each unit of the equipment is 6 persons. Customers arrive in Poisson fashion at a rate of 15 per hour.

(a) Determine the distribution for time between departures of each unit.

(b) When his assistant is on duty, he has been releasing units with 4, 5, and 6 customers on board with frequencies 0.2, 0.5, and 0.3, respectively. Find the mean and variance for time between departures under these conditions.

12. Patrons at a hamburger stand all arrive by automobile. A study reveals that each auto may contain 1, 2, 3, or 4 persons with frequencies 0.4, 0.2, 0.1, and 0.3, respectively. The autos arrive in Poisson fashion at a rate of 5 per hour. Customer service time is estimated to be uniformly distributed in the interval of 1–5 minutes.

(a) Find a transform for the distribution of number of customer arriving in time t.

(b) Express the probability generating function for the number of customers arriving during a service interval as a function of the Laplace transform of the service-time distribution.

13. Consider a plant maintenance department which must service the automatic transfer machines on a production line. The number of service calls per day can be described by a Poisson distribution with a mean of four. The time required to make a repair follows a negative exponential distribution with a mean of $\frac{1}{2}$ hour. The cost of a service call is estimated to be \$15 per hour spent in repair plus \$5 in paperwork to document the service. The life of the production line depends upon the vagaries of the marketplace. The present estimates are that the life of this line will be 200 days plus or minus 60 days.

(a) Find a transform for the distribution of daily repair time.

(b) Use the transform in part (a) to find the expected daily repair time.

(c) Find a transform for the distribution of daily costs.

(d) Estimate the mean and variance for the total cost of maintaining this equipment over the life of the production line.

14. Consider the random variable

$$y = 5x_1 + 10x_2 + 15x_3 + 32$$

where the x's are all exponentially distributed but each has a different mean: $E(x_1) = 3$; $E(x_2) = 2$; $E(x_3) = 6$. Find an appropriate transform for the probability that y exceeds r.

15. A wrestling bear is to be the feature attraction at the next campus fund raising party. The contribution given to the director of the selected charity I is made up from several random variables j each of which has a known mean and standard deviation μ_j and σ_j. He receives d per minute for each person present for the show. The bear will wrestle c rounds, the exact number depending upon the number of willing opponents g in the audience and the rounds k each can tolerate. The bear leaves when all opponents are exhausted. The length of each round is also a random variable t. Any campus organization may have w members present and the number of organizations represented is a random variable v.

(a) Use the tree diagram technique to find a functional relationship among the transforms of all the distributions involved. State all assumptions.

(b) Find the mean and variance for total income expressed as functions of the means and variances of the component variables.

16. A cut-rate calculator repair service is available on campus. Surplus transistors used by the repair service are purchased in bulk from a local supplier. Whenever the repair shop runs out, the supplier sends n transistors to replenish their stock. From past data, it is known that n has a mean of 1500 and a variance of 10,000. The price paid to the supplier C, an integer quantity, is uniformly distributed with a mean of 5 and variance of 6.67. The repair shop requires t minutes, distributed in exponential fashion with a mean of 20, to service one calculator. That service may require b transistors distributed binomially with parameters 0.2 and 10. The overhead rate is \$6 per hour. The shop charges for their services according to the degree to which they feel the customer can pay. Their rates are cost (including overhead) plus m, where m is normally distributed with a mean of 30 and variance of 16. Their inventory policy is to exhaust their supply before ordering more transistors. Customers arrive for service in Poisson fashion at a rate of 25 per day.

(a) Find a *transform* for the distribution of time spent serving one day's customers.

(b) Find the *mean* and *variance* for the number of transistors used per day.

(c) Find the *mean* and *variance* for charges per customer.

(d) Find the *mean* and *variance* for number of days required to dispose of one batch of transistors as received from the supplier.

17. Consider an inventory system which begins operating with Q units in stock. The inventory policy is one in which replacements are ordered immediately when customer demands take place, e.g., if customer A buys 3 units, a replenishment order of 3 is immediately initiated. Lead time is constant at 10 days. Time between customer orders is exponentially distributed by $f(t) = 2.5e^{-2.5t}$. The number of units r requested per customer is distributed by

u	$f(u)$
1	0.4
2	0.5
3	0.1

Customers who cannot be served from on-hand inventory are willing to backorder at a cost of \$5 per unit per year to the inventory system. Holding positive inventory costs \$2 per unit per year.

(a) Find an appropriate transform for the distribution of units demanded during a lead time.

(b) Find a transform for the distribution of units on order at any time t if lead time is Poisson distributed with mean 10.

18. Shippers the world over are about to launch an armada of ships in anticipation of the opening of a new canal. Their launching process is such that the time between ship arrivals at the entrance to the canal will follow the distribution

$$f(t) = 2 \exp(-t/0.5)$$

where t is measured in days. However, the canal cannot open until the canal commission can agree upon an equitable fee schedule. The time necessary to do that is estimated to be exponentially distributed with a mean of 10 days.

(a) Develop an appropriate transform for the distribution of the number of ships which will be in the queue when the canal opens.

(b) Estimate the probability that there will be two or more waiting when the canal opens.

19. In light of recent political developments, it is appropriate for the shippers of the world to reassess their policies concerning the use of the new canal. Now they feel that each launching event should send two ships on their way to the canal. The time between arrivals of *ship pairs* at the canal locks is now expected to be distributed by

$$f(t) = 3 \exp(-3t), \qquad t > 0$$

where t is measured in days. The time necessary to open the canal is estimated to be distributed by

$$g(t) = \binom{10}{t}(0.4)^t(0.6)^{10-t}, \qquad t = 0, 1, 2, \ldots$$

(a) Find an appropriate transform for the distribution of the number of ships which will be waiting at time t.

(b) Find the variance for the number waiting at the time the canal opens.

(c) Find the probability of more than 3 ships waiting when the canal opens.

(d) Find the probability of 9 ships arriving during a 3-day period.

20. In the analysis of certain waiting line problems the "busy period" for a system is sometimes described by the random variable $t = p + t_1 + t_2 +$

$\cdots + t_n$, where p is the processing time of a job distributed by $f(p)$ and n is Poisson distributed with a mean of λp.

(a) Prove that a proper functional equation involving the transform of the distribution of t, say $\mathscr{L}(h(t)) = H(s)$, is

$$H(s) = F(s + \lambda - \lambda H(s)), \qquad \text{where} \quad F(s) = \mathscr{L}(f(t))$$

(b) Use the equation in part (a) to find the mean and variance of the busy period as functions of the moments of the processing-time distribution.

CHAPTER 5

TRANSFORMS AND RANDOM EFFECTS IN LARGER SYSTEMS

The purpose of this chapter is to extend the probabilistic techniques introduced in the preceding chapters toward the analysis of increasingly realistic and hence more complex representations of systems. To achieve that goal, we will examine topics such as partial moments, multivariate distributions, and Markov processes. The common thread is the application of basic transform techniques to assist in the analysis and design of systems encompassing these properties.

Partial Moment Generating Functions

In many study areas, including statistical decision theory, Bayesian point estimation, and inventory theory, one frequently encounters terms we might refer to as "partial moments." For example, consider an inventory system in which demand over the lead time necessary to receive a replenishment order has the distribution $f(x)$, inventory position at the time of placing an order is

r, and the cost of denying service to a customer is d. The random variable y representing number of customers denied service during the lead time is

$$y = \begin{cases} 0, & r > x \\ x - r, & r \le x \end{cases}$$

The expected number of customers denied service is given by

$$E(y) = \int_r^\infty (x - r) f(x)\, dx = \int_r^\infty x f(x)\, dx - r \int_r^\infty f(x)\, dx \tag{5-1}$$

The expected cost of denying service is obviously

$$E(c) = dE(y)$$

The first integral in Eq. (5-1) is referred to as a partial expectation. It receives that name because the argument normally associated with the expected value of a random variable is integrated over only a portion of the range of definition for that random variable. If our cost function had been quadratic, the second partial moment would have been needed.

In order to standardize the notation, let us define the nth partial moment of x about the origin as

$$E_{-\infty}^{r}(x^n) = \int_{-\infty}^{r} x^n f(x)\, dx \tag{5-2}$$

The corresponding expression for discrete variables is given by

$$E_{-\infty}^{r}(m^n) = \sum_{-\infty}^{r} m^n p(m) \tag{5-3}$$

One can now express any desired partial expectation as a function of these standard forms. For example, the partial expectation of Eq. (5-1) can be written as

$$E_r^\infty(x) = \int_r^\infty x f(x)\, dx = E(x) - E_{-\infty}^{r}(x)$$

In the general case, if the partial expectation is defined over the interval $[a, b]$, we have

$$E_a^b(x^n) = E_{-\infty}^{b}(x^n) - E_{-\infty}^{a}(x^n) \tag{5-4}$$

Although one can often carry out this integration or summation in straightforward fashion, the calculations may become extremely complicated. What we now seek are ways to exploit our knowledge of transforms to facilitate the evaluation of the nth partial moments of x about the origin.[1]

[1] For a more complete discussion of the various ways to calculate partial moments, see Winkler *et al.* [34].

It is customary to define the partial moment generating function as an exponential transform:

$$M_{-\infty}^{r}(s) = E_{-\infty}^{r}(e^{sx}) = \int_{-\infty}^{r} e^{sx} f(x)\, dx \tag{5-5}$$

If one wishes to use the Laplace transform as a generating function, the corresponding partial transform is

$$\mathscr{L}_0^r(f(x)) = F_0^r(s) = \int_0^r e^{-sx} f(x)\, dx \tag{5-6}$$

The partial Fourier transform could be used in a similar fashion. For discrete variables one could employ either the geometric of z-transform. For example, the partial geometric transform for nonnegative variables would be

$$G_0^r(z) = \sum_{m=0}^{r} z^m p(m) \tag{5-7}$$

Although any of the transforms could be used for evaluating the partial moments, we will use the Laplace and geometric transforms to illustrate the general techniques. The use of other transforms is left as an exercise for the reader.

Let us begin with the Laplace transform as a partial moment generating function. Following the same general approach used to generate the total moments of a distribution, consider first the derivatives of the partial transform.

$$\frac{d^n}{ds^n} \mathscr{L}_0^r(f(x)) = \int_0^r (-1)^n x^n e^{-sx} f(x)\, dx$$

The value of such a derivative at $s = 0$ yields the nth partial moment. That is,

$$E_0^r(x^n) = (-1)^n \frac{d^n}{ds^n} \mathscr{L}_0^r(f(x)) \bigg|_{s=0} \tag{5-8}$$

The complete generating function can be developed from the limit of the partial transform as the range of integration approaches infinity. That is,

$$\mathscr{L}(f(x)) = \lim_{r\to\infty} \mathscr{L}_0^r(f(x)) = \lim_{r\to\infty} \int_0^r e^{-sx} f(x)\, dx \tag{5-9}$$

Similarly, the complete moments can be viewed as limiting cases of the partial moments:

$$E(x^n) = \lim_{r\to\infty} (-1)^n \frac{d^n}{ds^n} \mathscr{L}_0^r(f(x)) \bigg|_{s=0} \tag{5-10}$$

In order to determine the moments about any arbitrary point a, we can utilize our knowledge of the shifting property of transforms. By definition the nth partial moment about a is

$$E[(x-a)^n] = \int_0^r (x-a)^n f(x)\, dx \tag{5-11}$$

As was discussed in Chapter 4, we can view the argument in the expectation as a new random variable expressed as a function of the variable x. That is, let

$$y = h(x) = (x-a)$$

Then define the partial generating function for $g(y)$, the distribution of y, as

$$\mathscr{L}_0^r(g(y)) = E_0^r(e^{-sy}) = \int_0^r e^{-s(x-a)} f(x)\, dx$$

Rewriting this expression as a function of the partial transform for $f(x)$, we have

$$\mathscr{L}_0^r(g(y)) = e^{sa} \mathscr{L}_0^r(f(x)) \tag{5-12}$$

In order to emphasize the moment generating aspect of this function, let us adopt the notation

$$L_0^r((x-a)) = \mathscr{L}_0^r(g(y)) \tag{5-13}$$

where it is understood that the function being transformed is the distribution of $y = (x-a)$, not the function $(x-a)$.

By taking derivatives of the L-function, we are led to

$$\begin{aligned} E_0^r[(x-a)^n] &= (-1)^n \frac{d^n}{ds^n} L_0^r \bigg|_{s=0} \\ &= (-1)^n \frac{d^n}{ds^n} [e^{sa} \mathscr{L}_0^r(f(x))] \bigg|_{s=0} \end{aligned} \tag{5-14}$$

It is obvious that complete moments about the point $x = a$ can be developed as a special case of this computation. That is,

$$E[(x-a)^n] = (-1)^n \frac{d^n}{ds^n} [e^{sa} \mathscr{L}(f(x))] \bigg|_{s=0} \tag{5-15}$$

Furthermore, the most important moment about a point other than the origin is the second moment about the mean, which for the total distribution is the variance. It therefore follows that the partial and complete variance expressions can be written as

$$E_0^r[(x-\mu)^2] = \frac{d^2}{ds^2} [e^{s\mu} \mathscr{L}_0^r(f(x))] \bigg|_{s=0} \tag{5-16}$$

and

$$\operatorname{var}(x) = \frac{d^2}{ds^2}[e^{s\mu}\mathscr{L}(f(x))]\bigg|_{s=0} \tag{5-17}$$

where μ is the expected value of x.

The same type of analysis can be extended to the discrete case in which the partial transform is the geometric transform of Eq. (5-7). As before, the complete generating function can be viewed as the limit of the partial function

$$G(z) = \lim_{r\to\infty} G_0^r(z) = \lim_{r\to\infty} \sum_{m=0}^{r} z^m p(m) \tag{5-18}$$

The partial factorial moments are generated in straightforward fashion by evaluating the derivatives of the partial transform at $z = 1$. That is,

$$E_0^r(y^{(n)}) = \frac{d^n}{dz^n} G_0^r(z)\bigg|_{z=1} \tag{5-19}$$

which follows directly from

$$\frac{d^n}{dz^n}\left[\sum_{m=0}^{r} z^m p(m)\right] = \sum_{m=0}^{r} m(m-1)\cdots(m-n+1)z^{m-n}p(m)$$

Applying the shifting theorem in order to evaluate moments about an arbitrary point a, we obtain

$$G_0^r(z \mid y = x - a) = E_0^r(z^{x-a}) = z^{-a}G_0^r(z) \tag{5-20}$$

where

$$G_0^r(z) = \sum_{m=0}^{r} z^m p(m)$$

From this it follows that the nth partial factorial moment about a can be obtained from

$$E_0^r[(x-a)^{(n)}] = \frac{d^n}{dz^n}[z^{-a}G_0^r(z)]\bigg|_{z=1} \tag{5-21}$$

An interesting computation develops if we consider the limiting case for the second factorial moment about the mean of the complete distribution. In this instance we make the observation that we can write

$$E[(x-\mu)^{2}] = E[(x-\mu)^2 - (x-\mu)]$$

since

$$E(x - \mu) = E(x) - \mu = 0$$

We can now observe that the factorial equivalent is

$$(x - \mu)^{(2)} = (x - \mu)(x - \mu - 1)$$

It therefore follows that

$$\text{var}\,(x) = \frac{d^2}{dz^2}[z^{-\mu}G(z)]\bigg|_{z=1} \tag{5-22}$$

The equivalent partial second moment might require that the mean of the distribution be replaced by the partial expectation of x.

Example Consider a computer facility which accepts jobs for processing during the first shift on the basis of the processing time required for those jobs. Any program with an anticipated run time greater than a minutes is denied first shift service. Those jobs are run during the second shift. A study of processing times for all jobs submitted during the past month indicates that the processing-time distribution can be described by a gamma distribution of the form

$$f(t) = \frac{t^{r-1}\exp(-t/b)}{\Gamma(r)b^r}, \qquad t \geq 0$$

The problem is to find the mean and variance for processing times of jobs run on the first shift for the special case of $a = 3$ minutes, $r = 2$, and $b = 0.5$.

We can begin by noting the relationship of partial moments to conditional expectations. The conditional distribution of service time given that service time is less than a minutes is given by the truncated distribution

$$h(t \mid t \leq a) = \frac{f(t)}{\left[\int_0^a f(t)\,dt\right]}, \qquad t \leq a$$

where $f(t)$ is the distribution for all service times. The expected value of this truncated distribution is the conditional mean. That is,

$$E(t \mid t \leq a) = \int_0^a th(t \mid t \leq a)\,dt = \int_0^a t\frac{f(t)\,dt}{F(a)}$$

where $F(a)$ represents the cumulative distribution for total service time evaluated at $t = a$. However, the term in the numerator is nothing more than the first partial moment of $f(t)$. The denominator could be viewed in similar fashion as the zeroth-order partial moment. It then follows that the conditional expectation can be written in general as a suitably normalized partial moment, namely,

$$E(t \mid t \leq a) = \frac{E_0^a(t)}{F(a)} = (-1)\frac{d}{ds}\frac{\mathscr{L}_0^a(f(t))}{F(a)}\bigg|_{s=0} \tag{5-23}$$

The nth conditional moment about zero is, in like manner,

$$E(t^n \mid t \leq a) = (-1)^n \frac{d^n}{ds^n}\frac{\mathscr{L}_0^a(f(t))}{F(a)} \tag{5-24}$$

By using the shifting theorem from Eq. (5-16), we find that the special case for conditional variance becomes

$$E[(t - \mu_c)^2 \mid t \leq a] = [F^{-1}(a)]\frac{d^2}{ds^2}[e^{s\mu_c}\mathscr{L}_0^a(f(t))]\bigg|_{s=0} \tag{5-25}$$

where μ_c is the conditional expected value of t.

Returning now to the computer processing-time problem, the service-time distribution is given by

$$f(t) = 4te^{-2t}, \qquad t \geq 0$$

The partial transform for this distribution is

$$\begin{aligned}
\mathscr{L}_0^3(f(t)) &= \int_0^3 4te^{-(2+s)t}\,dt \\
&= \left[\frac{2}{s+2}\right]^2 \{1 - e^{-3(s+2)}[1 + 3(s+2)]\} \\
&= \left[\frac{2}{s+2}\right]^2 F(3 \mid s+2)
\end{aligned}$$

where the parameter $b^{-1} = 2$ in the gamma distribution has been replaced by $(s + 2)$ in

$$F(3 \mid s+2) = \int_0^3 f(t)\,dt\bigg|_{b^{-1}=s+2}$$

The expected processing time for first shift jobs is then

$$\begin{aligned}
E(t \mid t \leq 3) &= (-1)\frac{d}{ds}\frac{\mathscr{L}_0^3(f(t))}{F(3)}\bigg|_{s=0} \\
&= \frac{1 - 25e^{-6}}{1 - 7e^{-6}} \\
&= \frac{0.9375}{0.9825} = 0.954 \text{ minute}
\end{aligned}$$

The variance for first shift jobs is, from Eq. (5-25),

$$\operatorname{var}(t \mid t \le 3) = \frac{d^2}{ds^2}\left[e^{0.954s}\mathscr{L}_0^3(f(t))\right]\bigg|_{s=0} \div 0.9825$$

$$= \frac{0.4517}{0.9825} = 0.4597$$

From further study of this facility we know that when a job is submitted it must be logged in by a former programmer in the group who has been demoted. Because of the demotion, the worker has unilaterally decided to accept a maximum of 10 jobs per hour before discretely leaving his desk. At that time a secretary must fill in for the balance of the hour. Data on submission rates indicate that the number of jobs received in an hour can be described by a geometric distribution with parameter $p = 0.1$. We are now interested in finding the expected number of jobs each hour which will be logged by the disgruntled former programmer.

The proper measure of performance for this individual involves the first- and zeroth-order partial moments of $f(m)$. Let J represent the number of jobs this worker handles in 1 hour and m be the number submitted. Then

$$J = \begin{cases} m, & m \le 10 \\ 10, & m > 10 \end{cases}$$

from which

$$E(J) = \sum_{m=0}^{10} mf(m) + 10\sum_{m=11}^{\infty} f(m) = E_0^{10}(m) + 10[1 - E_0^{10}(m^0)] \tag{5-26}$$

In this case the partial generating function is

$$G_0^{10}(z) = \sum_{m=0}^{10} z^m(0.1)(0.9)^m = 0.1\sum_{m=0}^{10}(0.9z)^m$$

The summation contains a term of the form of a^x, which, from our early discussion of sum calculus, yields the indefinite sum

$$\sum a^x = \frac{a^x}{a-1}$$

It therefore follows that

$$G_0^{10}(z) = \frac{0.1(0.9z)^m}{0.9z - 1}\bigg|_{m=0}^{11} = \frac{0.1[(0.9z)^{11} - 1]}{0.9z - 1} \tag{5-27}$$

From the transform, the necessary moments are

$$E_0^{10}(m^0) = G_0^{10}(z)\Big|_{z=1} = \frac{0.1(0.314 - 1)}{-0.1} = 0.686$$

$$E_0^{10}(m) = \frac{d}{dz} G_0^{10}(z)\Big|_{z=1}$$

$$= \frac{1.1z^{10}(0.9)^{11}(0.9z - 1) - 0.09[(0.9z)^{11} - 1]}{(0.9z - 1)^2}\Bigg|_{z=1}$$

$$= \frac{-0.11(0.314) - 0.09(-0.686)}{0.01} = 2.72$$

Therefore

$$E(J) = 2.72 + 10(0.314) = 5.86 \text{ jobs/hour}$$

This leaves the secretary with an average of 3.14 jobs per hour as fill-in.

Multivariate Distributions

There are many system problems in which it is desirable to work with joint distributions with each of the random variables retaining its identity throughout the calculations. For example, consider the two-phase service system faced by student purchasers of athletic activity cards at some universities. They must first pay the university fees to obtain evidence that they are bona fide students then purchase their activity cards at a second facility. A possible model to use in this system is the so-called two-phase queueing model. The output from the model would give not only the probability distribution for total students in the system, but also the distribution for students waiting at each separate service facility.

Determination of the probability distributions in such multivariate problems is aided by the use of multivariate generating functions. For example, in the queueing illustration above we would be concerned with integer-valued random variables n_1 and n_2 with a joint distribution of the form

$$\Pr[n_1 = j, n_2 = k] = p(j, k)$$

For this distribution let us define a generating function, or geometric transform, that depends on two transform variables z_1 and z_2:

$$G_{1,2}(z_1, z_2) = \sum_{n_1=0}^{\infty} \sum_{n_2=0}^{\infty} z_1^{n_1} z_2^{n_2} p(n_1, n_2) \tag{5-28}$$

Such transforms will be called bivariate for short. The extension to joint distributions with larger numbers of variables is obvious. Furthermore, if the joint density is a continuous function, we could use an appropriate transform, such as the Laplace transform, to write the corresponding continuous transform as

$$F_{1,2}(s_1, s_2) = \int_0^\infty \int_0^\infty [\exp(-s_1 t_1 - s_2 t_2)] f(t_1 t_2)\, dt_1\, dt_2 \qquad (5\text{-}29)$$

Certain properties of these multivariate transforms are evident from their definitions. For example, the transforms for the marginal distributions can be obtained by appropriate substitution of terms. Since

$$p(n_1) = \sum_{n_2} p(n_1, n_2)$$

it follows that the transform for $p(n_1)$ is

$$G_1(z_1) = \sum_{n_1} z_1^{n_1} p(n_1) = \sum_{n_1} z_1^{n_1} \sum_{n_2} p(n_1, n_2) = \sum_{n_1} \sum_{n_2} z_1^{n_1} p(n_1, n_2)$$

But this expression looks exactly like Eq. (5-28) with the argument z_2 replaced by unity. The marginal transforms for integer-valued random variables are then

$$G_1(z_1) = G(z_1, z_2)\Big|_{z_2=1} = G(z_1, 1) \qquad (5\text{-}30)$$

and

$$G_2(z_2) = G(z_1, z_2)\Big|_{z_1=1} = G(1, z_2) \qquad (5\text{-}31)$$

The corresponding transforms for the continuous case are

$$F_1(s_1) = F(s_1, s_2)\Big|_{s_2=0} = F(s_1, 0) \qquad (5\text{-}32)$$

and

$$F_2(s_2) = F(s_1, s_2)\Big|_{s_1=0} = F(0, s_2) \qquad (5\text{-}33)$$

Proof of these assertions is left as an exercise for the reader.

The generating function for the sum of random variables can also be developed from the multivariate transform. If we define

$$m = n_1 + n_2$$

then the generating function for m can be written as

$$G_m(z) = E(z^m) = E(z^{n_1+n_2})$$

or

$$G_m(z) = \sum_{n_1} \sum_{n_2} z^{n_1+n_2} p(n_1, n_2)$$

But this expression looks like Eq. (5-28) in which the arguments z_1 and z_2 are replaced by a single argument z. That is,

$$G_m(z) = G_{1,2}(z_1, z_2)\Big|_{z_1=z_2=z} = G(z, z) \tag{5-34}$$

for the sum of two random variables with a known bivariate generating function. The obvious extension to the sum $t = t_1 + t_2$ of continuous variables yields

$$F_t(s) = F_{1,2}(s_1, s_2)\Big|_{s_1=s_2=s} = F_{1,2}(s, s) \tag{5-35}$$

In passing we also note that two variables x_1 and x_2 are independent if and only if

$$T_{1,2}(s_1, s_2) = T_1(s_1)T_2(s_2) \tag{5-36}$$

where T represents any of our usual transformations. This is a direct result of the requirement that the joint density of two independent random variables is formed by the product of the marginal densities.

Multivariate Moments

The techniques used to generate moments for distributions of single random variables with various transforms are easily extended to distributions of several variables. Consider, for example, the trivariate distribution $f(x, y, z)$. If it is necessary to find the rth moment of x, one could integrate the joint density as

$$E(x^r) = \int_z \int_y \int_x x^r f(x, y, z)\, dx\, dy\, dz \tag{5-37}$$

Similarly, one could find the joint moments from

$$E(x^a y^b z^c) = \int_z \int_y \int_x x^a y^b z^c f(x, y, z)\, dx\, dy\, dz \tag{5-38}$$

where a, b, and c are any nonnegative integers.

Now suppose that the joint moment generating function (exponential transform) is available. By definition

$$M(s_1, s_2, s_3) = \int_z \int_y \int_x [\exp(s_1 x + s_2 y + s_3 z)] f(x, y, z)\, dx\, dy\, dz \qquad (5\text{-}39)$$

The first partial derivative of M with respect to s_1 yields

$$\frac{\partial M}{\partial s_1} = \int_z \int_y \int_x x[\exp(s_1 x + s_2 y + s_3 z)] f(x, y, z)\, dx\, dy\, dz$$

By evaluating this function with all s_i variables set equal to zero, we find that

$$E(x) = \left.\frac{\partial M}{\partial s_1}\right|_{s_1 = s_2 = s_3 = 0} \qquad (5\text{-}40)$$

Similarly, if we take partials with respect to s_1, s_2, and s_3, we find that

$$\frac{\partial^3 M}{\partial s_1\, \partial s_2\, \partial s_3} = \int_x \int_y \int_z xyz[\exp(s_1 x + s_2 y + s_3 z)] f(x, y, z)\, dx\, dy\, dz$$

from which the joint first moment is

$$E(xyz) = \left.\frac{\partial^3 M}{\partial s_1\, \partial s_2\, \partial s_3}\right|_{s_1 = s_2 = s_3 = 0} \qquad (5\text{-}41)$$

The extension of the above results to higher order moments is obvious. If one wants $E(x^a y^b z^c)$, one can differentiate a times with respect to s_1, b times with respect to s_2, and c times with respect to s_3 and then set all s-values equal to zero. That is,

$$E(x^a y^b z^c) = \left.\frac{\partial^{a+b+c} M}{\partial^a s_1\, \partial^b s_2\, \partial^c s_3}\right|_{s_1 = s_2 = s_3 = 0} \qquad (5\text{-}42)$$

The same approach can be employed for other types of multivariate transforms. For example, appropriate joint factorial moments could be generated from the joint geometric transform by taking derivatives with respect to z_i and then evaluating with the z_i's equal to unity. Proof of this assertion is left as an exercise for the reader.

Multivariate Normal Functions

In many applications involving functions of random variables one may be interested in only the mean and variance of those functions. Furthermore, one may have available data from which the mean, variance, and perhaps correlation coefficients among the variables can be calculated. What we now seek are further ways to express the statistics of a function of several random variables as a function of statistics on its components. In the discussion which follows, the joint density is assumed to be multivariate normal. For a more complete discussion of these techniques, the reader is urged to consult Haugen [18, Chapter 3].

The random variables $x_1, x_2, \ldots, x_n$ are said to have a multivariate normal distribution if the joint probability density function is given by

$$f(x_1, x_2, \ldots, x_n) = \frac{1}{(2\pi)^{n/2}\sqrt{|\boldsymbol{\sigma}|}} \exp\left[-\tfrac{1}{2}(\mathbf{x} - \boldsymbol{\mu})\boldsymbol{\sigma}^{-1}(\mathbf{x} - \boldsymbol{\mu})'\right] \tag{5-43}$$

where

(i) $-\infty < x_i < \infty$
(ii) $\mathbf{x} = (x_1, x_2, \ldots, x_n)$
(iii) $\boldsymbol{\mu} = (\mu_1, \mu_2, \ldots, \mu_n)$
(iv) $\boldsymbol{\sigma} = \{\sigma_{ij}\}$ is the covariance matrix, i.e., a square matrix whose elements are the second moments of the variables.

The moment generating function for this distribution is given by[2]

$$M(s_1, s_2, \ldots, s_n) = \exp\left(\sum_{k=1}^{n} s_k \mu_k + \tfrac{1}{2}\sum_{i=1}^{n}\sum_{j=1}^{n} \sigma_{ij} s_i s_j\right) \tag{5-44}$$

For the special case of the bivariate normal this is often expressed as

$$M(s_1, s_2) = \exp\left[s_1\mu_1 + s_2\mu_2 + \tfrac{1}{2}(s_1^2\sigma_1^2 + 2\rho s_1 s_2 \sigma_1\sigma_2 + s_2^2\sigma_2^2)\right] \tag{5-45}$$

where $\rho = \sigma_{12}/\sigma_1\sigma_2$ is the correlation coefficient.

Let us now employ the bivariate normal moment generating function to develop expressions for commonly encountered functions of random variables. Consider, for example, the product

$$y = x_1 x_2 \tag{5-46}$$

From Eqs. (5-42) and (5-45) we can find

$$E(y) = \left.\frac{\partial^2 M}{\partial s_1\,\partial s_2}\right|_{s_1 = s_2 = 0} \qquad \text{and} \qquad E(y^2) = \left.\frac{\partial^4 M}{\partial s_1^2\,\partial s_2^2}\right|_{s_1 = s_2 = 0}$$

from which the variance can be calculated as

$$\sigma_y^2 = E(y^2) - [E(y)]^2$$

[2] See, for example, Laning and Battin [28, p. 75], where the characteristic function for the multivariate normal with zero means is discussed.

Taking derivatives, we obtain

$$\frac{\partial M}{\partial s_1} = (\mu_1 + s_1\sigma_1^2 + \sigma_{12}s_2) \exp(A)$$

where (A) represents all of the terms in the exponent of Eq. (5-45). The second partial is

$$\frac{\partial^2 M}{\partial s_1\, \partial s_2} = [\sigma_{12} + (\mu_1 + s_1\sigma_1^2 + s_2\sigma_{12})(\mu_2 + s_2\sigma_2^2 + s_1\sigma_{12})] \exp(A) \tag{5-47}$$

The fourth partial is

$$\begin{aligned}\frac{\partial^4 M}{\partial s_1^2\, \partial s_2^2} = \{&\sigma_1^2\sigma_2^2 + 2\sigma_{12}^2 + 4\sigma_{12}(\mu_1 + s_1\sigma_1^2 + s_2\sigma_{12})(\mu_2 + s_2\sigma_2^2 + s_1\sigma_{12}) \\ &+ \sigma_1^2(\mu_2 + s_2\sigma_2^2 + s_1\sigma_{12})^2 + \sigma_2^2(\mu_1 + s_1\sigma_1^2 + s_2\sigma_{12})^2 \\ &+ (\mu_1 + s_1\sigma_1^2 + s_2\sigma_{12})^2(\mu_2 + s_2\sigma_2^2 + s_1\sigma_{12})^2\} \exp(A)\end{aligned} \tag{5-48}$$

It follows that the expected value of the product $y = x_1 x_2$ is

$$E(y) = \left.\frac{\partial^2 M}{\partial s_1\, \partial s_2}\right|_{s_1 = s_2 = 0} = \mu_1\mu_2 + \sigma_{12} \tag{5-49}$$

The second moment of the product is

$$\begin{aligned}E(y^2) &= \left.\frac{\partial^4 M}{\partial s_1^2\, \partial s_2^2}\right|_{s_1 = s_2 = 0} \\ &= \{\sigma_1^2\sigma_2^2 + 2\sigma_{12}^2 + 4\mu_1\mu_2\sigma_{12} + \mu_2^2\sigma_1^2 + \mu_1^2\sigma_2^2 + \mu_1^2\mu_2^2\}\end{aligned} \tag{5-50}$$

The variance of the product can now be expressed as

$$\sigma_y^2 = E(y^2) - [E(y)]^2 = \sigma_1^2\sigma_2^2 + \mu_1^2\sigma_2^2 + \mu_2^2\sigma_1^2 + 2\mu_1\mu_2\sigma_{12} + \sigma_{12}^2 \tag{5-51}$$

In the event that x_1 and x_2 are independent, $\sigma_{12} = 0$. The mean and variance of the product of two random variables can then be expressed as simple functions of the means and variances of the components as

$$E(x_1 x_2) = \mu_1\mu_2 \tag{5-52}$$

$$\text{var}\,(x_1 x_2) = \sigma_1^2\sigma_2^2 + \mu_1^2\sigma_2^2 + \mu_2^2\sigma_1^2 \tag{5-53}$$

Moments for the square of a normally distributed random variable, the square root of a random variable, quadratic forms, and other combinations are developed in similar fashion. For details, the reader should consult Haugen [18, pp. 122–128].

Example A group of psychiatrists has established a counseling clinic for both married couples and single persons. Psychiatrist C handles only married couples, B handles only single males, and A handles only single females. The number of couples appearing for counseling each week follows a Poisson distribution with mean c. The distributions for single males and females are also Poisson with means b and a, respectively. The problem is to study the weekly counseling load at the clinic.

In order to maintain the mathematical distinction between males and females, let us introduce two transform variables z_1 and z_2. The generating functions for single males and single females are obtained from the familiar generating function for the Poisson distribution as

$$G_A(z_1) = e^{-a(1-z_1)} \tag{5-54}$$

and

$$G_B(z_2) = e^{-b(1-z_2)} \tag{5-55}$$

If we treat each couple as a single entity, the generating function for number of couples is, in like fashion,

$$G_C(z) = e^{-c(1-z)} \tag{5-56}$$

However, in order to develop statistics by sex, let us redefine the arrival of each couple as the joint event of one male and one female arriving. The joint distribution associated with each arrival event is then

$$p(n_1, n_2) = \begin{cases} 1 & \text{for} \quad n_1 = n_2 = 1 \\ 0 & \text{otherwise} \end{cases}$$

since all couples include one female and one male, by definition. The bivariate transform for this distribution is the very simple form

$$G_n(z_1, z_2) = z_1 z_2 \tag{5-57}$$

But the number of such joint events in a day is Poisson distributed. It therefore follows that the joint distribution for males and females generated by arriving couples has the transform

$$G_C(z_1, z_2) = e^{-c(1-z_1 z_2)} \tag{5-58}$$

by application of our earlier nesting arguments for compound distributions. The total number of persons processed by counselor C can be generated from Eq. (5-58) by recalling the relationship of Eq. (5-34). Since the total n is given by

$$n = n_1 + n_2$$

it follows that

$$G_n(z) = G_{1,2}(z, z) = G_C(z, z) = \exp[-c(1 - z^2)] \tag{5-59}$$

To develop appropriate generating functions for the entire clinic, let us assume that the arrival populations are independent. The bivariate clinic generating function is then

$$\begin{aligned} G(z_1, z_2) &= G_A(z_1)G_B(z_2)G_C(z_1, z_2) \\ &= \exp(-a - b + az_1 + bz_2 - c + cz_1z_2) \end{aligned} \tag{5-60}$$

The generating function for the number of women served by the clinic is

$$G_1(z_1) = G(z_1, z_2)\Big|_{z_2=1} = \exp[-(a + c)(1 - z_1)]$$

This is easily recognized as the transform of a Poisson distribution with a mean of $a + c$. In similar fashion we can show that the number of men served is Poisson distributed with a mean of $b + c$. However, the total number of persons served by the clinic, $x_1 + x_2$, is not Poisson. The transform for that distribution is, from Eqs. (5-34) and (5-60),

$$G(z) = G(z_1, z_2)\Big|_{z_1=z_2=z} = \exp[-(a + b)(1 - z) - c(1 - z^2)] \tag{5-61}$$

Appropriate statistics can now be developed by standard manipulation of these transforms.

Markov Chains

The Markov property can be viewed as a first-order generalization of the common probability modeling assumption of independent trials. In an experiment involving independent trials the random portion of the outcome of any trial is in no way dependent upon the outcomes of past trials. At the opposite end of the spectrum one might imagine an experiment in which the sequence of events is such that the entire past history of the experiment is needed to make probability statements about the next trial in the sequence. With Markov processes we are interested in moving one step from the trivial independent-trials assumption toward the more realistic dependence features of most stochastic processes. Here we assume one stage dependence in which the outcome at any trial depends upon the outcome of the immediately preceding trial but is independent of all others. For an excellent discussion of Markov analysis the reader is urged to consult the text by Howard [21].

In order to construct a Markov model of a system, one needs to understand the basic concepts of "state" and "state transition." The term "state" implies all we need to know to describe the system at any instant. To simplify the model, the number of variables in the state description should be kept to a minimum. In a waiting line model, for example, the state at time t might mean simply the number of customers present at that instant. In an inventory system the state might mean the number of units on hand, etc. State transitions imply that changes of state have occurred. It is often easiest to associate these transitions with time as we attempt to model dynamic systems. However, time is not an inherent part of the description of Markov processes since state transitions can be triggered by events other than the passage of time.

One can model both continuous and discrete parameter processes using the Markov assumption. However, for the purpose of illustration in this chapter, we will limit our attention to discrete parameter processes only. Those Markov processes that are discrete in both time and space are referred to as Markov chains.

Let us use a single real number to describe the outcome of a trial in a series of experiments. The random variable associated with trial or transition n then takes on a particular value k with some frequency:

$$\Pr(x_n = k) = f_k^n \tag{5-62}$$

The Markov property assures us that a description of the trajectory expected on the nth trial can be reduced to

$$\Pr[x_n = j \mid x_1 = k_1, x_2 = k_2, \ldots, x_{n-1} = i] = \Pr[x_n = j \mid x_{n-1} = i] \tag{5-63}$$

These transition probabilities are the heart of the Markov model. If transition probabilities are dependent on both the stage (trial number) and state of the system, we will write

$$\Pr[x_n = j \mid x_{n-1} = i] = p_{ij}(n) \tag{5-64}$$

for the probability of moving from state i to state j at the nth stage or transition. If the system is stationary, these transition probabilities are no longer a function of the stage number. In that case the notation can be shortened to

$$p_{ij}(n) = p_{ij} \qquad \text{for all} \quad n$$

Unless otherwise stated, all of our future discussion will assume stationarity.

Assuming that trials are defined by discrete increments, we can build up conditional probability distributions for transitions over long intervals from those over short intervals by means of the Chapman–Kolmogorov equation.

Let p_{ij}^n represent the probability of moving from state i to state j in n transitions. Then

$$p_{ij}^n = \sum_k p_{ik}^r p_{kj}^{n-r} \tag{5-65}$$

One could verbalize this expression by stating that the system moves from state i to some intermediate state k in r transitions, then from state k to the final state j in the remaining $n - r$ transitions. The joint probabilities of these events, when summed over all possible intermediate states k, yield the n-stage transition probability p_{ij}^n.

The most important application of Eq. (5-65) occurs in the development of the basic recursion equations that describe the n-stage behavior of a Markov chain. If a system starts in state i at time zero, the probability that it will occupy state j at time n can be calculated in recursive fashion from the single-stage transition probabilities as

$$p_{ij}^n = \sum_k p_{ik} p_{kj}^{n-1} \tag{5-66}$$

This is merely a special case of Eq. (5-65) with $r = 1$.

Beginning with $n = 1$, we note that

$$p_{kj}^0 = \begin{cases} 1, & k = j \\ 0, & \text{elsewhere} \end{cases} \tag{5-67}$$

from which $p_{ij}^1 = p_{ij}$, as expected. For $n = 2$ we have

$$p_{ij}^2 = \sum_k p_{ik} p_{kj}^1$$

which can be used to determine the three-stage transition probabilities, and so forth.

The mechanics of performing this calculation can be simplified by employing matrix notation. Let us construct a matrix the elements of which are the single-stage transition probabilities p_{ij}. That is,

$$\mathbf{P} = [p_{ij}] = \begin{bmatrix} p_{11} & p_{12} & \cdots & p_{1m} \\ p_{21} & p_{22} & \cdots & p_{2m} \\ \vdots & & & \\ p_{m1} & p_{m2} & \cdots & p_{mm} \end{bmatrix} \tag{5-68}$$

for a system with m possible states. Note the characteristics of this matrix.

1. The matrix is square ($m \times m$) since we must account for all possible transitions from any feasible initial state to any feasible final state.
2. All elements in the matrix are nonnegative since they represent probability values.
3. The sum of element values for each row in the matrix is unity since

each row represents a conditional probability distribution. That is, element j in row i represents the conditional probability of finding the system in state j on the next trial given that it currently occupies state i.

A matrix with these three characteristics is referred to as a "stochastic matrix."

The probability distribution for the state of the system resulting from trial n can be represented by the vector

$$\mathbf{V}(n) = (v_1(n), v_2(n), \ldots, v_m(n)) \tag{5-69}$$

To simplify the notation, we will sometimes drop the argument n in the elements of the vector. The ith element v_i then represents the probability that the system occupies state i at the particular trial of interest. The trial number will still be identified in the functional notation for the vector $\mathbf{V}(n)$. The matrix $\mathbf{P}$ together with an initial probability distribution $\mathbf{V}(0)$ is sufficient to completely specify a Markov chain. That is to say that one can develop complete probability statements about the "trajectory," or sequence of states occupied by the process, on the basis of the information contained in $\mathbf{P}$ and $\mathbf{V}(0)$.

Consider the calculation of state probabilities for a particular trajectory of a Markov process. Suppose that we want to know the probability that the system will start in state i_0, move from there to i_1, then to i_2, and so forth through n transitions. The probability of this particular sequence or *chain* of events is given by

$$\Pr(x_0 = i_0, x_1 = i_1, \ldots, x_n = i_n) = v_{i_0}(0)p_{i_0 i_1}p_{i_1 i_2} \cdots p_{i_{n-1} i_n} \tag{5-70}$$

as can be seen from the simple one-step, conditional probability statements.

If we are interested in the probability that the system might occupy any particular state j after one transition given that we know the initial state probabilities, we could find it from

$$\Pr(x_1 = j) = \sum_i \Pr(x_0 = i) \Pr(x_1 = j \mid x_0 = i)$$

or

$$v_j(1) = v_1(0)p_{1j} + v_2(0)p_{2j} + \cdots + v_m(0)p_{mj} \tag{5-71}$$

But this equation is nothing more that the inner product of the vector $\mathbf{V}(0)$ and column j of the $\mathbf{P}$ matrix. Since Eq. (5-71) is an expression for finding the system in some particular state j after one transition, it seems reasonable to suppose that the vector $\mathbf{V}(1)$, containing the entire state probability distribution, could be obtained by multiplying the vector $\mathbf{V}(0)$ by each column of the $\mathbf{P}$ matrix in turn. Symbolically we have the matrix equation

$$\mathbf{V}(1) = \mathbf{V}(0)\mathbf{P} \tag{5-72}$$

If one takes the position that $\mathbf{V}(1)$ represents current knowledge, the probability distribution after the second transition should be

$$\mathbf{V}(2) = \mathbf{V}(1)\mathbf{P} \tag{5-73}$$

Since $\mathbf{V}(1)$ is itself a function of $\mathbf{V}(0)$ and $\mathbf{P}$, this expression can also be written as

$$\mathbf{V}(2) = \mathbf{V}(0)\mathbf{P}\mathbf{P} = \mathbf{V}(0)\mathbf{P}^2$$

The general recursion formula is obviously

$$\mathbf{V}(n) = \mathbf{V}(n-1)\mathbf{P} \tag{5-74}$$

which can also be written as

$$\mathbf{V}(n) = \mathbf{V}(0)\mathbf{P}^n \tag{5-75}$$

Equations (5-74) and (5-75) form the basis for the rest of our analysis of Markov chains. In some sense it is convenient to view the matrix $\mathbf{P}$ as an operator which takes a probability distribution or vector at time n and transforms it into a new distribution at time $n + 1$. The matrix $\mathbf{P}$ is then a system which takes $\mathbf{V}(n)$ as input and yields $\mathbf{V}(n + 1)$ as output. Diagrammatically, we have

$$\mathbf{V}(n) \rightarrow \boxed{\quad P \quad} \rightarrow \mathbf{V}(n+1) \tag{5-76}$$

The elements of the matrix $\mathbf{P}^n$ contained in Equation (5-75) are the n-stage transition probabilities p_{ij}^n first introduced in Eqs. (5-65) and (5-66). The multiplication of matrices merely gives us a convenient method of bookkeeping for developing the conditional probability of moving from state i to state j in n transitions.

It should be noted that Eq. (5-74) is applicable to any Markov chain with a finite number of states. One does not need to know the theoretical niceties of irreducibility, ergodic properties, periodicity, etc. to perform the indicated calculation. However, the reader wishing to understand more of the theoretical structure should consult one of the many excellent references on Markov chains which are available. There one may explore such terms in depth. For our purposes we will take the pragmatic view that since Eq. (5-74) seems to work well, we will use it with little heed given to the vast literature that explores the implications of such manipulations.

It should be noted in passing that many system models are based on so-called steady-state solutions. This means that the system has run sufficiently long that the effects of initial conditions are no longer evident. Furthermore, the system has stabilized in the sense that the state probability distribution no longer changes with time.

Obviously some systems will never reach steady state. For example, periodic phenomena may be present or limiting distributions may be a function of starting state, which would make the idea of a limiting vector meaningless. However, one can operate as though a steady-state solution does exist and seek to find a vector $\mathbf{V}$ such that

$$\mathbf{V}(n+1) = \mathbf{V}(n) = \mathbf{V} \qquad \text{or} \qquad \mathbf{V} = \mathbf{VP} \tag{5-77}$$

In some sense steady state has been reached when the vector $\mathbf{V}(n)$ operated on by the matrix $\mathbf{P}$ again yields the vector $\mathbf{V}(n)$. The resulting m equations in the m unknown elements v_i are homogeneous. To obtain a solution, one must add the condition

$$\sum_{i=1}^{m} v_i = 1 \tag{5-78}$$

which must be satisfied by all probability distributions.

The beauty of this calculation is that anytime the equations can be solved the solution is meaningful. If the process is not one with a single closed set of states, which is required for steady-state solutions, the attempted solution of those equations will lead to an indeterminant set of equations rather than an incorrect solution. If the process is periodic, the solution obtained represents the relative frequency with which each state is occupied over a long history of transitions rather than a time steady-state vector. Let us define the distribution $\mathbf{V}$ as a measure of frequency of state occupancy during a long history; then if a solution to Eqs. (5-77) and (5-78) exists, it is valid.

Example Consider the problem faced by a meteorologist in attempting to provide long-range forecasts for a local area. For modeling purposes he is willing to classify daytime weather phenomena into three groups and assume that the system is Markovian. Sunny days are those in which more than one-half of the hourly observations show less than 0.6 cloud cover and no precipitation. Cloudy days are those in which more than one-half of the hourly observations show over 0.6 cloud cover but no precipitation. Rainy days are those in which precipitation is recorded. Data are available for 100 sunny days, 200 cloudy days, and 300 rainy days and their immediate successors.

State for test day	Observations for succeeding day		
	Sunny	Cloudy	Rainy
Sunny	30	20	50
Cloudy	80	60	60
Rainy	60	180	60

When reduced to relative frequency terms, we have the following transition matrix:

$$\mathbf{P} = \begin{pmatrix} 0.3 & 0.2 & 0.5 \\ 0.4 & 0.3 & 0.3 \\ 0.2 & 0.6 & 0.2 \end{pmatrix}$$

where state 1 = sunny, state 2 = cloudy, and state 3 = rainy.

Suppose that the weather is sunny today and that the meteorologist is preparing forecasts to cover the next three days. The current state of the weather can be described by the vector

$$\mathbf{V}(0) = (1, 0, 0)$$

Probability estimates for tomorrow's weather can be obtained from

$$\mathbf{V}(1) = \mathbf{V}(0)\mathbf{P} = (0.3, 0.2, 0.5)$$

From this he would predict a 50% chance of rain tomorrow. Continuing ahead to a 48-hour forecast, the probabilities are

$$\mathbf{V}(2) = \mathbf{V}(1)\mathbf{P} = (0.27, 0.42, 0.31)$$

From this he might forecast a 30% chance of rain for the second day. In like manner the 72-hour forecast could be obtained from

$$\mathbf{V}(3) = \mathbf{V}(2)\mathbf{P} = (0.311, 0.366, 0.323)$$

This time the chance of rain is > 32%.

Forecasts can be made for as many days in advance as the meteorologist has courage to trust the calculations. As each new day begins, he can examine his current state, as determined by yesterday's weather, to prepare his revised forecast. In addition he can also estimate the long-run frequency for each type of day. The steady-state solution is obtained from

$$\mathbf{V} = \mathbf{VP}$$

or

$$(v_1, v_2, v_3) = (v_1, v_2, v_3)\mathbf{P} \qquad \text{and} \qquad v_1 + v_2 + v_3 = 1$$

Rewriting, we have

$$\begin{aligned} v_1 &= 0.3v_1 + 0.4v_2 + 0.2v_3 \\ v_2 &= 0.2v_1 + 0.3v_2 + 0.6v_3 \\ v_3 &= 0.5v_1 + 0.3v_2 + 0.2v_3 \\ 1 &= \quad v_1 + \quad v_2 + \quad v_3 \end{aligned}$$

Since we now have four equations in three unknowns, we can eliminate one homogeneous equation from the set, say the first one. Solving the remaining equations simultaneously,

$$0.2v_1 - 0.7v_2 + 0.6v_3 = 0$$
$$0.5v_1 + 0.3v_2 - 0.8v_3 = 0$$
$$v_1 + v_2 + v_3 = 1$$

we obtain

$$v_1 = 0.304, \qquad v_2 = 0.368, \qquad v_3 = 0.328$$

for the steady-state solution. This indicates that over long time periods only about 30% of the days will be sunny.

Transform Solutions for Markov Chains

Our knowledge of transform techniques can be put to good use in finding solutions for the Markov chain problem. The key is to recognize that Eq. (5-74) can be viewed as a simple difference equation in n. The fact that the terms in the equation involve vectors and matrices poses no conceptual difficulty. The transform of a matrix is another matrix formed by transforming every element in the original matrix. These observations, coupled with adherence to the basic rules that distinguish matrix algebra from simple algebra, should permit us to find a complete transform solution to the Markov chain problem.

Rewriting Eq. (5-74), we have

$$\mathbf{V}(n + 1) = \mathbf{V}(n)\mathbf{P} \tag{5-79}$$

In accordance with our usual procedure for solving difference equations, let us form the subsidiary equation by taking the geometric transform of each side of Eq. (5-79). From the table of transform pairs we know that shifting is accomplished by

$$G(f(n + 1)) = z^{-1}(G(f(n)) - f(0))$$

Furthermore, the matrix **P**, being simply a collection of real numbers, can be treated as a constant in forming the subsidiary equation. It therefore follows that the subsidiary equation for Eq. (5-79) is

$$\mathbf{G}(\mathbf{V}(n + 1)) = \mathbf{G}(\mathbf{V}(n)\mathbf{P})$$

or

$$z^{-1}(\mathbf{G}(\mathbf{V}(n)) - \mathbf{V}(0)) = \mathbf{G}(\mathbf{V}(n))\mathbf{P} \tag{5-80}$$

where $\mathbf{G}(\mathbf{V}(n))$ is the single unknown. The elements of the vector $\mathbf{G}(\mathbf{V}(n))$

consist of a set of m transforms, which, when inverted, give the desired probability distribution after any number of transitions n.

Solving Eq. (5-80) for the unknown vector yields

$$\mathbf{G}(\mathbf{V}(n)) = \mathbf{V}(0)[\mathbf{I} - z\mathbf{P}]^{-1} \tag{5-81}$$

Note that Eq. (5-81) is a matrix equation. Here $\mathbf{I}$ represents the $m \times m$ identity matrix and the negative exponent represents the matrix inversion operation. Recovery of the vector $\mathbf{V}(n)$ can be accomplished by taking the inverse transform of every element in the matrix $[\mathbf{I} - z\mathbf{P}]^{-1}$ and weighting the result by the initial probability vector.

It is instructive to compare the form of Eq. (5-81) with the expression for $\mathbf{V}(n)$ presented in Eq. (5-75). Transforming Eq. (5-75), we obtain

$$\mathbf{G}(\mathbf{V}(n)) = \mathbf{G}(\mathbf{V}(0)\mathbf{P}^n) = \mathbf{V}(0)\mathbf{G}(\mathbf{P}^n) \tag{5-82}$$

It therefore follows that

$$[\mathbf{I} - z\mathbf{P}]^{-1} = \mathbf{G}(\mathbf{P}^n) \tag{5-83}$$

As a result of this calculation we can find both the state probability distribution after an arbitrary number n of transitions and the set of conditional probabilities p_{ij}^n for moving from state i to state j in n transitions.

Properties of the Inverse Transform

Howard makes several observations about the closed form expression for $\mathbf{P}^n$ which will be obtained from the inversion of Eq. (5-83) [21, p. 62]. If the process consists of a single closed set, meaning that any state can be reached from any other state at some point in time, we can write

$$\mathbf{P}^n = \mathbf{S} + \mathbf{T}(n) \tag{5-84}$$

where the rows of $\mathbf{S}$ are identical. The stochastic matrix $\mathbf{S}$ arises from inverting a term in the transform of the form $1/(1 - z)$. This is equivalent to saying that the matrix $\mathbf{P}$ has an eigenvalue of 1, as can be proven for all stochastic matrices. The matrix $\mathbf{S}$ represents the steady-state solution for the system. The matrix $\mathbf{T}(n)$, which may be factorable into a sum of several differential matrices (i.e., matrices whose rows sum to zero), represents the transient portion of the solution. The terms of $\mathbf{T}(n)$ will always have coefficients of the form c^n, nc^n, n^2c^n, etc., where the absolute value of c never exceeds unity. In some sense $\mathbf{T}(n)$ can be viewed as causing perturbations about the steady-state solution $\mathbf{S}$.

The form of Eq. (5-84) is equally appropriate for those processes that may not be so well behaved as the simple single closed set chains with which we are used to working. If the process does have a single closed set, the matrix **S** will have every row identical. The elements of that repeated row will be the same values one would obtain from the steady-state calculation of Eq. (5-77). If, on the other hand, the process is multichained in that it has more than a single closed set of states, each row of **S** may be unique. In that case the proper interpretation of row i of the **S** matrix is that of a limiting conditional distribution. The element s_{ij} of that matrix represents the conditional probability of finding the system in state j, after a large number of transitions, given that it started in state i. Attempting to find a solution for such a system by using Eq. (5-77) would lead to an indeterminant set of equations. The transforms thus provide very powerful techniques for finding closed form expression for probability distributions and high-order transitions. These techniques are independent of the particular theoretical class of Markov chain which might evolve in a modeling situation.

Example The local liquor dealer has collected information about bourbon-drinking habits of his regular clientele, i.e., those who visit the store every Friday night to purchase a bottle of weekend cheer. He stocks only two competing brands, B_1 and B_2. As each customer is served he asks him what brand he drank last week and duly notes what brand the customer is purchasing this week. This brand-switching raw data collected over the past two weeks shows the following counts:

Number of customers	Behavior
7	Remained with B_1
28	Switched from B_1 to B_2
22	Remained with B_2
33	Switched from B_2 to B_1

The problem is to establish some measure of the relative demand for each product.

Assuming that the Markov model is an adequate description of customer behavior, let us convert the raw data to the following transition matrix for states B_1 and B_2:

$$\mathbf{P} = \begin{bmatrix} 7/35 & 28/35 \\ 33/55 & 22/55 \end{bmatrix} = \begin{bmatrix} 0.2 & 0.8 \\ 0.6 & 0.4 \end{bmatrix}$$

Applying the transform relationship of Eq. (5-83), we obtain

$$(\mathbf{I} - z\mathbf{P}) = \begin{pmatrix} 1 - 0.2z & -0.8z \\ -0.6z & 1 - 0.4z \end{pmatrix}$$

A matrix this small is most easily inverted by using the adjoint and determinant, namely

$$\mathbf{A}^{-1} = \frac{\text{adj } \mathbf{A}}{|\mathbf{A}|}$$

from which

$$\begin{aligned}(\mathbf{I} - z\mathbf{P})^{-1} &= \frac{1}{1 - 0.6z - 0.4z^2}\begin{pmatrix} 1 - 0.4z & 0.8z \\ 0.6z & 1 - 0.2z \end{pmatrix} \\ &= \frac{1}{(1 - z)(1 + 0.4z)}\begin{pmatrix} 1 - 0.4z & 0.8z \\ 0.6z & 1 - 0.2z \end{pmatrix}\end{aligned}$$

By partial fraction expansion this reduces to

$$(\mathbf{I} - z\mathbf{P})^{-1} = \frac{1}{1 - z}\begin{pmatrix} 3/7 & 4/7 \\ 3/7 & 4/7 \end{pmatrix} + \frac{1}{1 + 0.4z}\begin{pmatrix} 4/7 & -4/7 \\ -3/7 & 3/7 \end{pmatrix}$$

From the table of transform pairs the inverse of this expression is

$$\mathbf{P}^n = \begin{pmatrix} 3/7 & 4/7 \\ 3/7 & 4/7 \end{pmatrix} + (-0.4)^n \begin{pmatrix} 4/7 & -4/7 \\ -3/7 & 3/7 \end{pmatrix}$$

Finally, the distribution by brand can be developed from

$$\mathbf{V}(n) = \mathbf{V}(0)\mathbf{P}^n$$

If this model is a reasonable picture of customer purchasing habits, the dealer can expect the long-range demand for brand B_1 to be 3/7 and for brand B_2 to be 4/7 of his total business. If, for example, his customer population is 350, he can expect to sell an average of 150 bottles of B_1 and 200 bottles of B_2 each week. If he wants to predict the purchasing behavior of one of his customers who has just purchased brand B_1, the probability that this customer will also buy brand B_1 on his nth future purchase is

$$p_{11}^n = \tfrac{3}{7} + (-0.4)^n(\tfrac{4}{7})$$

Tracing this customer's predicted loyalty to brand B_1, we then have

$$v_1(0) = 1, \quad v_1(1) = 0.2, \quad v_1(2) = 0.52, \quad v_1(3) = 0.392, \quad v_1(4) = 0.4432$$

Numerical Inversion of Transforms

Throughout the preceding discussions there remains the haunting spector that in many real-world problems the compact expressions we have developed by theoretical arguments may prove to be intractable if we seek to invert the resulting transforms. So far we have avoided the issue by asserting that we can always recover the necessary moments or by suggesting that one only needs extensive tables and an uncanny insight to find the inverse. We will now seek to correct this shortcoming by suggesting certain numerical methods which can be used to invert some commonly encountered forms.

Compound Poisson Processes

A commonly recurring model in a great many applications, including inventory control, telephone traffic, accident damage, and fire prediction, is that of the compound Poisson process. This model arises when we consider a random sum of random variables in which the number of terms in the sum is Poisson distributed. That is, y is a compound Poisson variable when

$$y = x_1 + x_2 + \cdots + x_n$$

n follows the Poisson distribution and the x_i are identically distributed by some distribution $f(x)$.

From our previous work on the random sum of random variables, we know that the geometric transform for the distribution of y is given by

$$G_y(z) = G_n(G_x(z)) = \exp\left[-\mu(1 - G_x(z)\right] \tag{5-85}$$

where μ is the mean of the Poisson distribution. For an arbitrary distribution $f(x)$

$$G_x(z) = \sum_{j=0}^{\infty} f_j z^j \tag{5-86}$$

In practice the distribution for x can be approximated by a finite probability set. The numerical method for inverting Eq. (5-85) which follows is due to Van Landingham and Shariq [33].

The key to the numerical technique is to recognize that the transform for a compound Poisson process can be expressed as a product of transforms of the form $\exp(a_i z^i)$. That is,

$$\begin{aligned} G_y(z) &\approx \exp\left[-\mu + \mu(f_0 + f_1 z + f_2 z^2 + \cdots + f_m z^m)\right] \\ &= \exp(a_0 + a_1 z + a_2 z^2 + \cdots + a_m z^m) = \prod_{j=0}^{m} \exp(a_j z^j) \end{aligned} \tag{5-87}$$

Here it is assumed that $Pr(x > m)$ is negligible.

When written in this form, one can conclude from Borel's theorem that the sequence $h(y)$ must be the m-fold convolution of the sequences whose transforms are of the form $\exp(a_i z^i)$. The solution procedure then depends on the following theorem [33, p. 51]:

Theorem If $G(z) = \sum_{n=0}^{\infty} f_n z^n$, for $|z| < 1$, is the generating function for the probability set $\{f_n\}$, then $G(z^k)$ is the geometric transform for the related sequence of positive numbers $\{\nabla \sum_{m=0}^{[n/k]} f_m\}$, i.e.,

$$G(z^k) = g\left\{\nabla \sum_{m=0}^{[n/k]} f_m\right\} \tag{5-88}$$

where $[\cdot]$ represents "the integer part of" and ∇ is the backward difference $[\nabla F(n) = F(n) - F(n-1)]$.

The lower case g is used in Eq. (5-88) to distinguish the geometric transform of the term in braces from the transform of the f_n sequence denoted by $G(z)$.

The impact of this theorem is that we now have a means of expressing each factor of the product in Eq. (5-87) as a well-defined sequence. By power series expansion we know that

$$G(z) = \exp(a_i z) = \sum_{n=0}^{\infty} \frac{(a_i z)^n}{n!} \tag{5-89}$$

Combining this observation with the results of the theorem, we obtain

$$\exp(a_i z^i) = G\left\{\nabla \sum_{m=0}^{[n/i]} \frac{a_i^m}{m!}\right\} \tag{5-90}$$

But

$$\nabla \sum_{m=0}^{[n/i]} \frac{a_i^m}{m!} = \begin{cases} \dfrac{a_i^{n/i}}{(n/i)!}, & \text{for} \quad n = 0 \pmod{i} \\ 0, & \text{for} \quad n \neq 0 \pmod{i} \end{cases} \tag{5-91}$$

Although the sequence in Eq. (5-91) is not necessarily a probability set, it is a valid inverse for $\exp(a_i z^i)$. Furthermore, application of the convolution properties of the geometric transform is not limited to probability sequences alone. Any two sequences can be so convolved. It then follows that a compound Poisson distribution can be represented by the m-fold convolution:

$$h(y = n) = \{\exp[\mu(f_0 - 1)]\}\left[\left\{\frac{(\mu f_1)^n}{n!}\right\} * \begin{Bmatrix} \dfrac{(\mu f_2)^{n/2}}{(n/2)!}, & n \quad \text{even} \\ 0, & n \quad \text{odd} \end{Bmatrix}\right.$$
$$\left. * \cdots * \begin{Bmatrix} \dfrac{(\mu f_m)^{n/m}}{(n/m)!}, & n = 0 \pmod{m} \\ 0, & n \neq 0 \pmod{m} \end{Bmatrix}\right] \tag{5-92}$$

Computational Algorithm

As a practical matter, one must eventually truncate the sequences being convolved in Eq. (5-92) in order to obtain numerical solutions. However, the truncation of the sequences at some fixed number of terms N does not affect the results of the convolution for all $n \leq N$. To illustrate this point, suppose that the infinite sequences $\{a_i\}$ and $\{b_i\}$ are to be convolved. Let us truncate each sequence at $N = 3$, that is, we will set all terms with subscript higher than three equal to zero. The convolution $c = a * b$ will have terms

$$\begin{aligned} c_0 &= a_0 b_0 \\ c_1 &= a_0 b_1 + a_1 b_0 \\ c_2 &= a_0 b_2 + a_1 b_1 + a_2 b_0 \\ c_3 &= a_0 b_3 + a_1 b_2 + a_2 b_1 + a_3 b_0 \\ c_4 &= (a_0 b_4) + a_1 b_3 + a_2 b_2 + a_3 b_1 + (a_4 b_0) \end{aligned} \tag{5-93}$$

where the terms in parentheses will be zero for the truncated problem. Note that c_0, c_1, c_2, and c_3 are exact expressions for both the truncated and nontruncated cases. There is zero error caused by truncating at $N = 3$ (leaving four terms in each sequence) for all c_i where $i \leq 3$. Only those terms in the c-sequence of order higher than c_3 will be affected by the truncation.

This is an extremely powerful observation in a probability modeling context. This says that if we truncate each sequence in Eq. (5-92) to have the same number of terms N, convolving those truncated sequences will give us an error-free sequence for the first N terms of the compound Poisson distribution. If we are only interested in low-order probabilities such as Pr $(y \leq j)$, where $j < N$, we have an exact solution. Furthermore, we can always establish the amount of the distribution being ignored by subtracting the sum of the terms up to $h(N)$ in the convolution from unity. This also suggests a stopping rule for the algorithm. One can program a tolerance limit t. After each term $h(n)$ for the convolution is determined, compare the sum to the desired tolerance. When the inequality

$$\sum_{i=0}^{n} h(i) \geq 1 - t \tag{5-94}$$

is satisfied, the distribution has been determined to the desired level of accuracy. The availability of large-scale digital computers makes this approach feasible for most compound Poisson distributions one is likely to encounter in practical problems.

Numerical Inversion of Laplace Transforms

There are also techniques available for the numerical inversion of Laplace transforms. Most of them are based upon the idea of replacing an integral by a finite sum. This approximation then reduces the solution of the integral equation of the Laplace transform to the solution of a system of linear algebraic equations. Techniques for establishing the coefficients in the algebraic equations come under the title of numerical quadrature. They are based upon a fundamental understanding of Legendre polynomials. The procedures are still in a rather primative stage of development, requiring a mix of strategies, a strong knowledge of analysis, and considerable intuition to successfully apply them. They are beyond the scope of this text. However, interested readers can refer to the text by Bellman, Calaba, and Lockett [3] for a basic discussion of the state of the art.

Problems

1. Use Eq. (5-17) to find the variance for the following distributions:

(a) $f(x) = be^{-bx}, \quad x \geq 0.$

(b) $f(x) = \dfrac{x^{a-1}e^{-x/b}}{\Gamma(a)b^a}, \quad x \geq 0.$

(c) $f(x) = (b - a)^{-1}, \quad a \leq x \leq b.$

2. Use Eq. (5-22) to find the variance for the following distributions:

(a) $p(j) = \binom{n}{j}p^j(1 - p)^{n-j}, \quad j = 0, 1, \ldots, n.$

(b) $p(j) = \dfrac{a^j e^{-a}}{j!}, \quad j = 0, 1, 2, \ldots, \infty.$

(c) $p(j) = p(1 - p)^j, \quad j = 0, 1, 2, \ldots, \infty.$

3. Consider a single-period inventory system in which 20 items are purchased for resale at a price of \$100 each. The retail price is to be \$150 each. Any items left in inventory at the end of the month are discounted for sale at \$90 each to clear the stock. If a customer cannot be served during the month from the initial order of stock, management has decreed that they will obtain the item for their customer from a competitor at a price of \$170 even though it will be sold to their own customer at \$150. Demand is believed to be Poisson distributed with a mean of 15 items.

(a) Calculate the expected number of items purchased from a competitor.

(b) Find the expected number of units sold from stock.

(c) Find the variance for income from the end-of-month discount sale.

4. The output from the clippers in a plywood plant consists of two grades of veneer. Any sheets that exceed 8 feet are classed as grade A. Any that are less than or equal to 8 feet are classed as grade B. Past data indicate that the total population of veneer output is approximately uniformly distributed between 7 and 13 feet. Use both partial generating functions and direct calculation to solve the following problems:

(a) Find the average length of grade A veneer.

(b) Find the variance for grade B veneer.

5. A spray painter on an assembly line paints door panels at a rate of 15 per hour. After being painted they are inspected and classified into three groups; reject, repair, and satisfactory. Past data indicate that 10% will be rejected and 20% will require repair.

(a) Develop the generating function for the trinomial distribution that will yield the probability distribution for the joint event (x_0, x_1, x_2), where x_i represents the number of occurrences of event i in 1 hour's production.

(b) Use the generating function to find the distribution for the sum of rejects and repairs per hour.

(c) Find the joint mean and joint variance for the trinomial distribution.

6. Prove that the joint factorial moments of a distribution can be developed from the joint geometric transform.

7. Prove that Eq. (5-45) is valid.

8. Use the moment generating function for the normal distribution to prove that the mean and variance for the square of a normally distributed random variable are given by

$$\mu_y = \mu_x^2 + \sigma_x^2, \qquad \sigma_y^2 = 4\mu_x^2\sigma_x^2 + 2\sigma_x^4$$

where $y = x^2$ and x is normal.

9. The range of a certain radar antenna used for air traffic control is believed to be normally distributed with a mean of 40 miles and a variance of 16. Find the mean and variance of the circular area swept by this antenna.

10. Consider the quadratic form

$$y = ax^2 + bx + c$$

where x is normally distributed.

(a) Find the moment generating function for y.

(b) Find the mean of y expressed as a function of the moments of x.

(c) Find the variance of y expressed as a function of moments of x.

11. Consider the function

$$y = x_1 x_2^2 + x_1^2 x_3^3 + x_1 x_2 x_3$$

where x_1, x_2, and x_3 are normally distributed with known means, variances, and covariances. Find the expected value of y.

12. Consider the clinic counseling problem represented by Eq. (5-60).

(a) Find the probability that 1 male and 3 females will be counseled in 1 week.

(b) Find the mean and variance for number of persons served by the clinic over a 4-week period.

(c) Suppose that each person served pays the clinic according to his ability. The income per patient is uniformly distributed in the range of 10–20 dollars. Find the mean and variance for total cash received.

13. Consider the problem faced by a bear hunter firing at a running bear. He has n rounds of ammunition available. If the previous shot was a hit, the hunter has a relatively high probability of repeating the hit on the next shot. If the previous shot was a miss, he must adjust his aim and consequently has a lower probability of making a hit on the next shot. Field data show that the first shot will hit 10% of the time. Subsequently, a hit will follow another hit 80% of the time and a hit will follow a miss 40% of the time.

(a) Model the system as a stationary Markov process.

(b) Find a closed form expression for the probability of scoring a hit with the jth round.

(c) Suppose that at least 2 hits are required to kill the bear. Find the probability that the bear will escape if the hunter has $n = 5$ rounds to fire.

14. Find a closed form expression for $\mathbf{P}^n$ where

$$\mathbf{P} = \begin{pmatrix} 0 & 0 & 1 \\ 0.3 & 0.2 & 0.5 \\ 1 & 0 & 0 \end{pmatrix}$$

15. A part-time employment service is open for business on Monday, Wednesday, and Friday each week. The service is staffed by persons with limited skills. The state of the system is given by X_n, the number of persons available for hire at the beginning of business day n. The maximum number available is 3. On Monday an incoming request will match the skills of this agency with probability 0.8. On Wednesday this drops to 0.6, and on Friday to 0.4. Replacements for those hired out are obtained by telephoning members of the part-time pool each business day up to the limit of 3. The probability distribution for number of available replacements per business day is

$$g(r) = \begin{cases} 0.6, & r = 0 \\ 0.3, & r = 1 \\ 0.1, & r = 2 \end{cases}$$

All replacements occur at the start of the business day. Customer demands on day i are distributed by $f_i(d)$, where

$$f_{\mathrm{M}}(d) = \begin{cases} 0.5, & d = 0, \\ 0.3, & d = 1, \\ 0.2, & d = 2, \end{cases} \qquad f_{\mathrm{W}}(d) = \begin{cases} 0.4, & d = 0, \\ 0.3, & d = 1, \\ 0.3, & d = 2 \end{cases}$$

$$f_{\mathrm{F}}(d) = \begin{cases} 0.1, & d = 1 \\ 0.4, & d = 2 \\ 0.5, & d = 3 \end{cases}$$

$$\mathrm{M} = \text{Monday}, \ \mathrm{W} = \text{Wednesday}, \ \mathrm{F} = \text{Friday}.$$

(a) Model the business as a Markov chain. Fill in the appropriate numbers in all transition matrices.

(b) Show how to find the probability of having 3 potential hirees left at the end of the day on Friday if they started Monday before telephoning with 2 available.

(c) If it exists, describe how one could find the steady-state distribution for number of hirees available on Monday, Wednesday, or Friday. Explain all calculations.

Note: Throughout this problem set up the appropriate matrix equations but it is not necessary to complete the numerical calculations.

16. An aviation school has a single aircraft available for charter. Scheduling is handled by an electronic data processing system. The status of the aircraft is recorded only when it changes states. The datum recorded is $X_n = 0$ if the nth change of state puts it flying, and $X_n = 1$ if the nth change puts it in the repair shop. Shop scheduling is arranged so that work is never started on an aircraft needing repair, even though it is classed in the repair state, until they are certain that a customer will be waiting for it when repairs are completed. When the aircraft ceases flying it is equally likely to be because of a breakdown or lack of customers.

(a) Construct a single-stage transition matrix to describe the information system.

(b) Find the probability that the aircraft will enter flying status at a transition chosen at random after the system has been running for many hours, given that it entered flying status at time zero.

(c) Find an expression for $\mathbf{P}^n$.

17. Consider two aircraft A and B approaching simultaneously from the east for landing on parallel east–west runways. Their desired flight paths are separated by 1 mile. However, because of instrument errors, pilot techniques, wind shear, etc., the actual position of aircraft i at any particular point in time is a random variable described by the vector (x_{1i}, x_{2i}, x_{3i}). The component x_{1i} represents distance east of the runway threshold, x_{2i} represents distance north or south of a line midway between runway centerlines, and x_{3i} represents altitude above the surface. All first and second moments for the distributions of the nominal position coordinates at the outer marker are

$$\mu_A = (25{,}000;\ 2000;\ 2500) \qquad \mu_B = (24{,}000;\ -500;\ 2400)$$
$$\sigma_A^2 = (70{,}000;\ 121;\ 2500), \qquad \sigma_B^2 = (121;\ 250{,}000;\ 1600)$$

Develop a model and explain how you would use it to estimate the probability that separation at the outer marker will be less than 1000 ft. State all assumptions.

18. The supply sergeant at an Air Force Base has made a careful study of requisitions for WAC charts from base personnel. When a request for charts comes in it may ask for one or more charts. An appropriate distribution for number of charts requested x seems to be

$$f(x) = 0.8(0.2)^{x-1}, \qquad x = 1, 2, \ldots$$

The number of requisitions per day appears to be Poisson distributed with a mean of 4. The resupply cycle is 5 days long.

(a) Find the geometric transform for number of WAC charts needed during one resupply cycle.

(b) Use numerical methods to find the probability that more than 3 charts will be requested in 1 day.

19. Employ numerical methods to evaluate the first three terms of the probability distribution whose transform is given by

$$G_y(z) = \exp \frac{-5(0.6 - 0.6z)}{1 - 0.6z}$$

20. Write a computer program to evaluate numerically the distributions for compound Poisson processes to any prespecified level of completeness, e.g., the first n terms of the distribution. Check the program by resolving Problems 18 and 19.

CHAPTER 6

APPLICATIONS FROM THE THEORY OF QUEUES

The purpose of this chapter is to illustrate how one can make extensive use of the basic techniques discussed in Chapters 3–5 to find solutions to problems in more complex stochastic environments. Models from the field of queueing theory offer many opportunities to test our knowledge. By exploring these special applications it is hoped that the reader may gain a fuller understanding of the use of transform techniques in probability modeling.

The Queueing Problem

Although the total area of application is extremely large, we will illustrate the queueing problem in terms of waiting line systems involving customers seeking service. Any such system can be characterized by three major features:

1. There is an input of customers to the system who seek some type of service.

2. Customers may be forced to wait in a queue until they can be served. Service takes place according to some well-defined policy, the simplest of which is first come, first served.

3. A service mechanism exists by which means customers are removed from the system.

Conceptually all one requires to completely specify a queueing system is a description of the input population, the queue discipline to be used, and the properties of the service mechanism. The goal is to develop a model which will give some measure of the number of customers present in the waiting line at time t and some measure of the delay experienced by a customer passing through the system. It is the uncertainty present in such systems which makes the problem a challenging one. The best characterization of a system one may be able to accomplish is to describe the number of service demands and the performance of the service mechanism in terms of probability distributions. The problem is then to use these input and service distributions, together with an announced queue discipline, to develop probability statements concerning line length and delay times.

It takes very little imagination to deduce that there is virtually a limitless number of variations one could encounter in describing such systems. The vast array of literature on the subject attests to that observation. However, as with many such topics, there is a solid core of basic prototype models and techniques which are useful for a surprising range of applications. We will investigate a few of these core items, particularly those that lend themselves nicely to the use of transform techniques.

In order to communicate concisely to others the exact nature of the assumptions underlying the models being presented, a reasonably standard set of notation has evolved. For example, the Operations Research Society of America recommends that the following shortened notation be adopted for describing queueing systems:

(arrival process/service process/number of parallel services)

When necessary an appendage may be added of the form

.../limit on the number in the system/number in the source/queue discipline)

When such an appendage is omitted it is assumed that there is no limit to the number that can be in the system, the customer population is infinite, and the queue discipline is first come, first served. For example, among the problems which we will discuss are the following:

(M/M/1) Markovian input and service with one server

(M/G/1) Markovian input, general service distribution, and one server

Queues are not inherently Markovian. However, those systems most widely discussed in the literature seek to preserve that property where possible. The importance of the Markov assumption in this context cannot be overemphasized. The major change from our previous discussion is that the single-stage dependence introduced in the last chapter for discrete time systems will now be carried forward to continuous time systems. The Markov property will enable us to write equations for the probability of finding a system in any state at the time of transition as a function of its state at the immediately preceding transition. This calculation is so powerful that we will attempt to make the system look as if it is Markovian even when it is not. This is the tactic pursued when we introduce the idea of imbedded Markov chains to model the general service case.

The Pure Birth Process

The Poisson Process

Let us begin our study of queues by examining the Poisson process as an arrival or birth process with the following properties:

1. The probability of a single arrival during a small interval of time is $\lambda\,\Delta t$, where

$$\lim_{\Delta t \to 0} \frac{\lambda\,\Delta t}{\Delta t} = \lambda \tag{6-1}$$

2. The probability of more than a single arrival during a small interval of time is $O(\Delta t)$, where

$$\lim_{\Delta t \to 0} \frac{O(\Delta t)}{\Delta t} = 0 \tag{6-2}$$

(For a complete discussion of birth and death postulates the reader should consult the book by Feller [13, Chapter XVII].)

Loosely speaking, we are postulating an arrival mechanism in which customers arrive singly, the probability of an arrival is proportional to elapsed time, and time intervals can be selected which are sufficiently small to negate the possibility of more than one customer appearing. Furthermore, suppose that the state of the system is described by the number of customers generated up to time t. The associated probability function is

$p_n(t)$ = probability of n customers present at time t

The problem now is to develop a recursion relationship for $p_n(t)$ from the basic Markov assumptions.

The single-stage transition matrix describing changes between t and $t + \Delta t$ is given by

$$\mathbf{T} = \begin{array}{c|ccccc} & 0 & 1 & 2 & 3 & \cdots \\ \hline 0 & 1 - \lambda\,\Delta t - O(\Delta t) & \lambda\,\Delta t & O(\Delta t) & 0 & \cdots \\ 1 & 0 & 1 - \lambda\,\Delta t - O(\Delta t) & \lambda\,\Delta t & O(\Delta t) & \cdots \\ 2 & 0 & 0 & 1 - \lambda\,\Delta t - O(\Delta t) & \lambda\,\Delta t & \cdots \\ \vdots & & & & & \end{array}$$

The entries for this matrix are developed by arguing that a move from state i to $i + 1$ occurs if a single arrival takes place in Δt. The probability of that event is $\lambda\,\Delta t$. The system will remain in state i only if no arrivals occur in Δt. The probability of that event is 1 minus the probability of one or more arrivals, or $1 - \lambda\,\Delta t - O(\Delta t)$ and the probability of negative arrivals, or services, in the pure birth case is zero. Although with unlimited queues permitted this matrix is of infinite dimension, the same three nonzero entries appear in every row.

Let us now construct a vector $\mathbf{P}(t)$ to represent the probability distribution for customers present, or generated, up to time t:

$$\mathbf{P}(t) = (p_0(t), p_1(t), \ldots, p_n(t), \ldots)$$

The basic Markov assumption permits us to write the probability vector at time $t + \Delta t$ as

$$\mathbf{P}(t + \Delta t) = \mathbf{P}(t)\mathbf{T} \tag{6-3}$$

Carrying out the indicated multiplication column by column, we obtain

$$\begin{aligned} p_0(t + \Delta t) &= p_0(t)[1 - \lambda\,\Delta t - O(\Delta t)] \\ p_1(t + \Delta t) &= p_0(t)\lambda\,\Delta t + p_1(t)[1 - \lambda\,\Delta t - O(\Delta t)] \\ &\;\vdots \\ p_n(t + \Delta t) &= p_{n-2}(t)O(\Delta t) + p_{n-1}(t)\lambda\,\Delta t + p_n(t)[1 - \lambda\,\Delta t - O(\Delta t)] \end{aligned} \tag{6-4}$$

By rearranging terms and dividing by Δt, we can write this set of equations as

$$\begin{aligned} \frac{p_0(t + \Delta t) - p_0(t)}{\Delta t} &= -\lambda p_0(t) - p_0(t)\frac{O(\Delta t)}{\Delta t} \\ \frac{p_1(t + \Delta t) - p_1(t)}{\Delta t} &= \lambda p_0(t) - \lambda p_1(t) - p_1(t)\frac{O(\Delta t)}{\Delta t} \\ &\;\vdots \\ \frac{p_n(t + \Delta t) - p_n(t)}{\Delta t} &= \lambda p_{n-1}(t) - \lambda p_n(t) + [p_{n-2}(t) - p_n(t)]\frac{O(\Delta t)}{\Delta t} \end{aligned} \tag{6-5}$$

Let us now take the limit of both sides of these expressions as Δt goes to zero. The formal definition of the derivative of a function $f(t)$ is

$$\frac{df(t)}{dt} = \lim_{\Delta t \to 0} \frac{f(t + \Delta t) - f(t)}{\Delta t}$$

Furthermore, we know from the postulate of Eq. (6-2) that any term involving $O(\Delta t)$ in the numerator will go to zero. It therefore follows that

$$\frac{dp_0(t)}{dt} = -\lambda p_0(t), \qquad \frac{dp_n(t)}{dt} = \lambda p_{n-1}(t) - \lambda p_n(t), \quad n \geq 1 \tag{6-6}$$

This is a set of differential–difference equations. They are differential equations in terms of the continuous variable t and difference equations in terms of the discrete variable n.

Although it is easy to solve these equations by recursion, it is instructive to use our knowledge of transforms to achieve a solution. The techniques illustrated will provide a map for solving much more complicated sets of differential–difference equations which arise in the study of queues.

Transform Solution

To begin the solution for this infinite set of equations, let us define the geometric transform for $p_n(t)$ as

$$G(z, t) = \sum_{n=0}^{\infty} p_n(t)z^n \tag{6-7}$$

The transform is expressed as a function of both the discrete variable n and the continuous variable t, although at this stage of the calculation t is treated as a constant.

One might think of this as the transform of a probability distribution which exists at some particular time t. Since that distribution will change as t changes, we will retain the t-argument to permit the same functional notation to be used for the transform at any point in time. Note that the partial derivative of the $G(z, t)$ function with respect to t can be written as a sum of the derivatives of $p_n(t)$ as

$$\frac{\partial G(z, t)}{\partial t} = \frac{\partial}{\partial t} \sum_{n=0}^{\infty} p_n(t)z^n = \sum_{n=0}^{\infty} \frac{dp_n(t)}{dt} z^n \tag{6-8}$$

A common attack for an infinite system of difference equations is to attempt to reduce the system to a single equation in the geometric transform by appropriate multiplication and addition. Let us first multiply the nth equation in the set (6-6) by z^n as follows:

$$\begin{aligned}\frac{dp_0(t)}{dt}z^0 &= -\lambda p_0(t)z^0 \\ \frac{dp_1(t)}{dt}z^1 &= \lambda p_0(t)z^1 - \lambda p_1(t)z^1 \\ &\vdots \\ \frac{dp_n(t)}{dt}z^n &= \lambda p_{n-1}(t)z^n - \lambda p_n(t)z^n\end{aligned} \tag{6-9}$$

Adding these equations, we obtain the single expression

$$\begin{aligned}\sum_{n=0}^{\infty}\frac{dp_n(t)}{dt}z^n &= \lambda\sum_{n=0}^{\infty}p_n(t)z^{n+1} - \lambda\sum_{n=0}^{\infty}p_n(t)z^n \\ &= \lambda z\sum_{n=0}^{\infty}p_n(t)z^n - \lambda\sum_{n=0}^{\infty}p_n(t)z^n\end{aligned} \tag{6-10}$$

The expression on the left is nothing more than the partial derivative of the transform from Eq. (6-8). The terms on the right each involve a summation that can be recognized as the geometric transform of $p_n(t)$. The entire function can then be written as a partial differential equation in $G(z, t)$ as

$$\frac{\partial G(z, t)}{\partial t} = \lambda z G(z, t) - \lambda G(z, t) = \lambda(z - 1)G(z, t) \tag{6-11}$$

The next step is to use the Laplace transform to solve the resulting differential equation. At this stage the transform variable z is treated as a constant. Using our earlier results for the transform of the derivative of a function leads to the subsidiary equation

$$s\mathscr{L}(G(z, t)) - G(z, 0) = \lambda(z - 1)\mathscr{L}(G(z, t)) \tag{6-12}$$

Solving for the Laplace transform of $G(z, t)$, we have

$$\mathscr{L}(G(z, t)) = \frac{G(z, 0)}{s - \lambda(z - 1)} \tag{6-13}$$

The appearance of this expression is sometimes disturbing to students. We appear to have a Laplace transform of a geometric transform. The key of course is that the geometric transform is a function of both z and t. For purposes of calculation with the Laplace transform, the z-argument is suppressed and the function $G(z, t)$ is treated as one would treat any function $f(t)$ of the single variable t.

To recover the corresponding time functions, we must now invert the transforms in Eq. (6-13). The Laplace transform in this expression is of the form

$$\mathscr{L}(f(t)) = \frac{k}{s - a}$$

which from the table of transform pairs yields

$$f(t) = ke^{at}$$

The coefficient k is determined by the initial condition, which for our problem is the value of the geometric transform of $p_n(t)$ at $t = 0$. If we assume that no one is present at time zero, it follows that

$$k = \sum_{n=0}^{\infty} p_n(0)z^n = 1$$

By using these identities, the inverse Laplace transform for Eq. (6-13) is found to be

$$G(z, t) = e^{\lambda(z-1)t} \tag{6-14}$$

Note that this expression is still a transform, a geometric transform in this case. What now remains is to invert the geometric transform to recover the desired time-varying probability distribution. Expanding the right side of Eq. (6-14) as a power series in z, we obtain

$$G(z, t) = e^{-\lambda t} \sum_{n=0}^{\infty} \frac{(\lambda tz)^n}{n!} \tag{6-15}$$

Since the definition of the geometric transform for $p_n(t)$ is

$$G(z, t) = \sum_{n=0}^{\infty} p_n(t)z^n$$

it follows by comparing like powers of z that

$$p_n(t) = \frac{(\lambda t)^n e^{-\lambda t}}{n!} \tag{6-16}$$

This expression is of course the familiar Poisson distribution with a mean of λt. What started as a simple proportionality constant, in an expression for the probability of an event in a small time interval, turns out to be the mean rate at which events take place in a Poisson process. When one speaks of arrivals in a queueing system as occurring in Poisson fashion, the implication is that the system is Markovian with a constant probability of event occurrence.

Time between Events

The probability distribution for time between arrival events in a Poisson process can be developed as follows. If an arrival has just occurred, the time to the next arrival will be less than t from this point if and only if one or more arrivals occur in the interval $(0, t)$. But this expression for the cumulative interarrival distribution is already available from Eq. (6-16). Since the probability of one or more arrivals in $(0, t)$ is equivalent to 1 minus the probability of zero arrivals in that interval, it follows that

$$F(t) = 1 - e^{-\lambda t} \tag{6-17}$$

The density can be developed by taking the derivative of the cumulative function as

$$f(t) = \frac{dF(t)}{dt} = \lambda e^{-\lambda t} \tag{6-18}$$

This expression is the familiar negative exponential distribution with parameter λ. What we have just shown is that in a process in which the number of events in an interval t follows the Poisson distribution, the time between events will follow a related exponential distribution.

The negative exponential distribution is the only continuous distribution that exhibits a total lack of memory. This means that the distribution for time to the next event in a Poisson process is independent of the time elapsed since the previous event took place. This is the property required of a Markov process; the move to a new state depends only upon the current state and is independent of the events used to reach that state. Mathematically,

$$P(t > t_1 + t_2 \mid t > t_1) = P(t > t_2) \tag{6-19}$$

denotes a random variable t with the Markov property. Using the definition for conditional probability and the expression for the exponential density, we find that for this case

$$\begin{aligned} P(t > t_1 + t_2 \mid t > t_1) &= \frac{P(t > t_1 + t_2 \text{ and } t > t_1)}{P(t > t_1)} \\ &= \frac{\exp[-\lambda(t_1 + t_2)]}{\exp(-\lambda t_1)} = \exp(-\lambda t_2) \end{aligned} \tag{6-20}$$

It therefore follows that the time to the next arrival event in a Poisson process has the same distribution after the process has been running for any time t as it had at $t = 0$.

The conclusion is that the queueing system in which the time between events follows exponential distributions can be modeled as a Markov process. The "events" of interest in such systems are the arrival of customers and the completion of service for customers currently in the system. To say that a system has Poisson arrivals and exponential service times is equivalent to saying that both the arrival and service processes are Markovian, hence the standard notation (M/M/1).

The Generalized Birth Process

The first step in generalization from the simple Poisson process is to drop the assumption that the probability of an arrival or birth during $(t, t + \Delta t)$ is independent of the number of arrivals up to time t. Let us now assume that if n arrivals have already taken place, the probability of a single arrival in the next interval Δt equals $\lambda_n \, \Delta t$. Again the possibility of multiple arrivals is contained in a term $O(\Delta t)$ which approaches zero as Δt becomes small. Note that the time between transitions is exponentially distributed although the parameter of that distribution is now expressed as a function of the current state. The arrival process still exhibits a total lack of memory.

For the generalized pure birth process the equations developed in (6-6) are modified to become

$$\frac{dp_0(t)}{dt} = -\lambda_0 p_0(t), \qquad \frac{dp_n(t)}{dt} = \lambda_{n-1} p_{n-1}(t) - \lambda_n p_n(t), \quad n \geq 1 \qquad \text{(6-21)}$$

Explicit formulas for $p_n(t)$ require specification of values for λ_n and are uniquely determined by the initial conditions. Solutions for particular system configurations will be deferred to problem exercises.

The Birth–Death Process

The previous sections have discussed in some detail how the birth process can be used in a queueing system to describe the evolution of customer arrivals. It is now but a short step in logic to include the death process in a queueing model. " Deaths " in this context denote customer discharge. A death occurs when a customer who has been waiting for service has that service completed, or perhaps leaves for other reasons.

The death postulates must now be added to those of Eqs. (6-1) and (6-2) as modified for the generalized birth process. The death postulates are as follows:

1. The probability of a single service or death during a small interval of time Δt, in a system with $n > 0$ customers present, is $\mu_n \, \Delta t$, where

$$\lim_{\Delta t \to 0} \frac{\mu_n \, \Delta t}{\Delta t} = \mu_n \qquad \text{(6-22)}$$

When $n = 0$ no customer can be served, so $\mu_0 = 0$.

2. The probability of more than a single service during Δt is $O(\Delta t)$, where

$$\lim_{\Delta t \to 0} \frac{O(\Delta t)}{\Delta t} = 0 \tag{6-23}$$

From our previous discussion concerning the pure birth process it is obvious that these postulates lead to the conclusion that the time between service events, when the system is occupied, will follow a negative exponential distribution. Because both the interarrival and service time distributions are exponential, and hence memoryless, a queueing system modeled as a birth–death process is Markovian. The differential–difference equations can be developed as before by constructing a transition matrix for the interval $(t, t + \Delta t)$, then using the basic Markov recursion equation.

Transition Matrix

Let $p_{ij}(\Delta t)$ be the probability of moving from i to j customers in the system during an elapsed time of Δt. Let us further agree to ignore all terms $O(\Delta t)$ since we know they will drop in the limiting arguments. If the starting state is zero customers present, the system can move only to a higher order state or stay in state zero. Staying in state zero implies that no customers arrive in Δt or that one arrival and one service take place. The probability of those events is

$$p_{00}(\Delta t) = \lambda_0 \, \Delta t \, \mu_0 \, \Delta t + (1 - \lambda_0 \, \Delta t) \tag{6-24}$$

Moving from state zero to state one implies that the joint event of a single arrival and no services must occur. The probability is therefore

$$p_{01}(\Delta t) = \lambda_0 \, \Delta t(1 - \mu_0 \, \Delta t)$$

Moving from state zero to some state higher than one requires multiple arrivals. The probabilities for those events are zero:

$$p_{0j}(\Delta t) = 0, \qquad j > 1$$

Higher order rows for the transition matrix will each have three nonzero elements. The system can move from i to $i - 1$ present, remain in state i, or move from i to $i + 1$ in the interval from t to $t + \Delta t$. The probabilities for these events are

$$\begin{aligned} p_{i,\,i-1}(\Delta t) &= \mu_i \, \Delta t(1 - \lambda_i \, \Delta t) \\ p_{ii}(\Delta t) &= \lambda_i \mu_i \, \Delta t^2 + (1 - \lambda_i \, \Delta t)(1 - \mu_i \, \Delta t) \\ p_{i,\,i+1}(\Delta t) &= \lambda_i \, \Delta t(1 - \mu_i \, \Delta t) \end{aligned}$$

The complete transition matrix is then

$$
\mathbf{T} = \begin{array}{c|cccc}
 & 0 & 1 & 2 & \cdots \\
\hline
0 & \lambda_0\mu_0\,\Delta t^2 + (1 - \lambda_0\,\Delta t) & \lambda_0\,\Delta t(1 - \mu_0\,\Delta t) & 0 & \cdots \\
1 & \mu_1\,\Delta t(1 - \lambda_1\,\Delta t) & \lambda_1\mu_1\,\Delta t^2 + (1 - \lambda_1\,\Delta t)(1 - \mu_1\,\Delta t) & \lambda_1\,\Delta t(1 - \mu_1\,\Delta t) & \\
2 & 0 & \mu_2\,\Delta t(1 - \lambda_2\,\Delta t) & \lambda_2\mu_2\,\Delta t^2 + (1 - \lambda_2\,\Delta t)(1 - \mu_2\,\Delta t) & \\
\vdots & & & &
\end{array}
$$

Finding a Solution

The complete set of differential–difference equations for the general birth and death process follows from

$$\mathbf{P}(t + \Delta t) = \mathbf{P}(t)\mathbf{T} \tag{6-25}$$

By the same techniques used for the birth process we are led to

$$
\begin{aligned}
\frac{dp_0(t)}{dt} &= -\lambda_0 p_0(t) + \mu_1 p_1(t) \\
\frac{dp_n(t)}{dt} &= -(\lambda_n + \mu_n)p_n(t) + \lambda_{n-1}p_{n-1}(t) + \mu_{n+1}p_{n+1}(t), \qquad n > 0
\end{aligned} \tag{6-26}
$$

In theory, this set of equations can be solved by the same transform techniques used before. Namely

1. Define the transform

$$G(z, t) = \sum_{n=0}^{\infty} p_n(t)z^n$$

2. Combine equations to obtain a differential equation in $G(z, t)$.
3. Solve the differential equation for $G(z, t)$ by using Laplace transforms.
4. Invert $G(z, t)$ to obtain $p_n(t)$.

However, the execution of these steps for the model in question is not a simple task even for the most trivial specifications for λ_n and μ_n. Solutions are available for such restricted cases as $\lambda_n = \lambda$ and $\mu_n = \mu$ for all n, but even those results are so complex that the full time-varying solution is of little practical value.

As an aside, however, we should note that it is possible to obtain numerical solutions to such a set of differential–difference equations even for the special case of time-varying arrival and service rates. The key is to treat a truncated system in which we have only a finite number of equations in the set (6-26). The time-varying arrival and service rate patterns are divided into a series of step functions in which the rates are constant within a given step. The transient probability distribution within a given time step is determined by the numerical solution to the corresponding finite system of linear differential equations with constant coefficients. The output from this set of equations is then used to provide boundary conditions for the next time interval with its new arrival and service rates. Standard numerical methods are available in most computer software packages to solve the truncated case. Since most physical systems rarely have unlimited capacity, such an approach seems intuitively appealing. For an excellent discussion of this technique, the reader is urged to consult Koopman [25].

The Steady-State Solution

Because of the difficulties encountered in achieving the full time-varying solution to many queueing systems, it is common practice in most basic texts to focus on the so-called "steady-state" solution. The idea of the steady state is that the system has reached statistical equilibrium. One is assumed to be interested in the behavior of the system after it has been running for a long period of time. If the system is well behaved, the effects of all initial conditions should be gone. The equilibrium distribution is a stationary probability distribution which can be expressed as

$$p_n = \lim_{t \to \infty} p_n(t) \tag{6-27}$$

If one is only concerned with the statistical equilibrium properties of the system, either because of design considerations or because of the difficulty in finding a transient solution, it is not necessary to first develop the expression for $p_n(t)$. If the limiting probabilities exist, it follows that

$$\lim_{t \to 0} \frac{dp_n(t)}{dt} = 0 \tag{6-28}$$

This merely states that, since the probability distribution is no longer changing with time, the derivative of that function must be zero. We can now return to Eq. (6-26), set the derivatives equal to zero, and drop the time-dependent notation. The resulting set of equations for the steady-state solution is

$$\begin{aligned} 0 &= -\lambda_0 p_0 + \mu_1 p_1 \\ 0 &= -(\lambda_n + \mu_n)p_n + \lambda_{n-1} p_{n-1} + \mu_{n+1} p_{n+1}, \qquad n \geq 1 \end{aligned} \tag{6-29}$$

The problem has now been reduced to the simpler one of solving a set of ordinary difference equations.

The simplest way to obtain the general solution for this set of equations is by recursion. From the first equation we have

$$p_1 = \frac{\lambda_0}{\mu_1} p_0$$

Using that result in the second equation, we obtain

$$\begin{aligned} p_2 &= \frac{(\lambda_1 + \mu_1)p_1 - \lambda_0 p_0}{\mu_2} \\ &= \frac{(\lambda_1 + \mu_1)(\lambda_0/\mu_1)p_0 - \lambda_0 p_0}{\mu_2} = \frac{\lambda_0 \lambda_1}{\mu_1 \mu_2} p_0 \end{aligned}$$

In similar fashion the general case reduces to

$$p_n = \frac{\prod_{i=0}^{n-1} \lambda_i}{\prod_{i=1}^{n} \mu_i} p_0 \tag{6-30}$$

The unknown p_0 can be obtained from the normalized condition

$$\sum_{n=0}^{\infty} p_n = 1$$

from which

$$p_0 = \left[1 + \sum_{n=1}^{\infty} \left(\frac{\prod_{i=0}^{n-1} \lambda_i}{\prod_{i=1}^{n} \mu_i}\right)\right]^{-1} \tag{6-31}$$

If the summation in Eq. (6-31) does not converge, it is obvious that the value of p_0 and hence every value p_n will be zero. This simply means that there is no steady-state solution and that the system is growing without bound. It is important to note that for truncated systems the summation is finite, which guarantees convergence for all finite λ- and μ-values and hence the existence of a steady-state solution.

Special Configurations

Equations (6-30) and (6-31) are extremely powerful results. They can be used to construct equilibrium solutions for a great variety of system configurations. All that is required is to specify properly the set of values for $\{\lambda_n\}$ and $\{\mu_n\}$.

For example, consider the (M/M/1) model. Since the system has unlimited capacity and a constant service rate, we can specify

$$\lambda_n = \lambda \quad \text{for all} \quad n, \qquad \mu_n = \mu \quad \text{for all} \quad n > 0$$

Substituting into Eq. (6-31), we obtain

$$p_0 = \left[\sum_{n=0}^{\infty} \left(\frac{\lambda}{\mu}\right)^n\right]^{-1}$$

which will be nonzero only if $\lambda < \mu$. If that condition is met, the infinite sum is a geometric series which converges to

$$\sum_{n=0}^{\infty} \left(\frac{\lambda}{\mu}\right)^n = \frac{1}{1 - (\lambda/\mu)}$$

from which we obtain

$$p_0 = 1 - \frac{\lambda}{\mu} \qquad \text{and} \qquad p_n = \left(1 - \frac{\lambda}{\mu}\right)\left(\frac{\lambda}{\mu}\right)^n \tag{6-32}$$

If this system is truncated, meaning that it has some fixed capacity of k units, we can obtain the solution by defining

$$\lambda_n = \begin{cases} \lambda & \text{for} \quad n < k \\ 0 & \text{for} \quad n \geq k \end{cases}$$

This indicates that once the system is saturated no more customers are admitted, or equivalently the arrival rate goes to zero. We now have a finite sum, which we can evaluate from our knowledge of sum calculus:

$$\sum_{n=0}^{k} \left(\frac{\lambda}{\mu}\right)^n = \frac{(\lambda/\mu)^n}{(\lambda/\mu) - 1}\Bigg|_{n=0}^{k+1} = \frac{(\lambda/\mu)^{k+1} - 1}{(\lambda/\mu) - 1}$$

This leads to the distribution

$$p_n = \left(\frac{\lambda}{\mu}\right)^n \left(\frac{\lambda}{\mu} - 1\right) \Bigg/ \left[\left(\frac{\lambda}{\mu}\right)^{k+1} - 1\right] \tag{6-33}$$

for $n \leq k$. Note that this is a valid solution even when $\lambda > \mu$. Conceptually we can argue that even though the potential arrival rate exceeds the service rate, the act of truncation cuts the effective arrival rate to something less than the service rate to assure a stationary distribution.

Other steady-state models that fill the literature can be easily developed with the same type of reasoning. For example, consider a system with c parallel service channels, each operating in exponential fashion at a rate μ. Customers are assumed to form a single line and to be served in order by the first channel that becomes vacant. The service rate for the system as a whole then varies depending upon the number of channels occupied. We can tabulate the service rate and arrival rate as functions of number of customers present as follows:

n	μ_n	λ_n
0	0	λ
1	μ	λ
2	2μ	λ
3	3μ	λ
$\vdots$	$\vdots$	$\vdots$
c	$c\mu$	λ
$c + 1$	$c\mu$	λ
$\vdots$	$\vdots$	$\vdots$
$n > c$	$c\mu$	λ

Plugging these values into Eqs. (6-30) and (6-31) will give the solution for the classical multichannel queueing model. If the system is truncated at k customers, we simply set $\lambda_n = 0$ for all $n \geq k$ and proceed as before.

Comprehensive Example As a final illustration of the versatility of the steady-state model, consider the policy of a university maintenance department. The foreman has three mechanics who he can assign, if necessary, to repairing breakdowns among the school's 20 buses. However, since he also has other tasks which he wants the mechanics to perform, he is reluctant to commit them all at once to the repair task. His policy is to provide a single mechanic for buses as long as there are no more than 3 buses in the shop. When more than 3 but less than 7 buses are present, two mechanics are assigned. When 7 or more buses are down, all mechanics are required to work on bus repair. Furthermore, shop space is limited such that no more than 10 buses at a time can be accommodated. Breakdowns occurring when the shop is full are handled by a local auto dealer who provides a loaner vehicle while a bus is in his shop. An analysis of transportation department records indicates that each bus has a mean time between failures of 8 hours. Shop records indicate that a single mechanic will repair a breakdown in an average of 2 hours.

In order to use the birth–death model as an approximation for this system, let us assume that the probability distribution for time between failures on a bus and the distribution for repair times are both exponential. The arrival process is such that the total effective arrival rate when n buses are operating is $n\lambda$. If r of those n are in the shop, the arrival or breakdown rate goes to $(n - r)\lambda$. We can now construct the rate table for this truncated, finite-population, variable-server problem as follows:

n	μ_n	λ_n	n	μ_n	λ_n
0	0	20λ	6	2μ	14λ
1	μ	19λ	7	3μ	13λ
2	μ	18λ	8	3μ	12λ
3	μ	17λ	9	3μ	11λ
4	2μ	16λ	10	3μ	0
5	2μ	15λ	> 10	3μ	0

$$\lambda = 0.125/\text{hour}, \quad \mu = 0.50/\text{hour}$$

Particular numerical values for the steady-state distribution for the number of buses in the university garage under such a policy can now be obtained by applying Eqs. (6-30) and (6-31). Those calculations are left as an exercise for the reader.

Finding a Waiting-Time Distribution

The models of the previous sections have focused on developing probability distributions for the number of customers, including those currently in service, present in a queueing system. Once those distributions are known it is a relatively simple task to use transform techniques to develop any moments of the distributions that might be needed to evaluate performance. The one major item still missing is some measure of the delay time experienced by a customer seeking service. The task in this section is to develop a strategy for calculating a probability distribution for delay.

Let us consider again the equilibrium solution for simple (M/M/1) queue described by Eq. (6-32). If one takes the position of a customer arriving at random at some point in time after the system has reached steady state, his total delay time will depend upon the number of customers preceding him. If he happens to arrive when the system is empty, his delay time is simply the service time he requires. However, if one customer already occupies the system, his delay time is the sum of the remaining service time for that customer and his own service time. In similar fashion we can argue that with n customers present at the time our reference customer arrives, he must wait for the duration of service t_i for each of the n customers plus his own service time t_0.

Let the arriving customer's waiting time, conditioned on n customers present, be given by the random variable

$$t_{w|n} = t_0 + t_1 + t_2 + \cdots + t_n \tag{6-34}$$

If we assume statistical independence, the conditional waiting-time distribution is the convolution of the $n+1$ distributions $f(t_i)$. However, since this system has exponentially distributed service times with the convenient property of forgetfulness, the conditional waiting-time distribution is simply the $(n+1)$-fold convolution of

$$f(t) = \mu e^{-\mu t} \tag{6-35}$$

From our knowledge of Laplace transforms it therefore follows that the transform for the conditional waiting-time distribution, which we might treat as a conditional transform, is

$$\mathscr{L}(f(t_{w\,|\,n})) = F(s\,|\,n) = [\mathscr{L}(f(t))]^{n+1} = \left(\frac{\mu}{\mu + s}\right)^{n+1} \tag{6-36}$$

To simplify the final calculation, note that the number of customers present and hence the number of terms in the summation is a random variable. This means that we are again faced with a random sum of random variables. The distribution for n is the geometric distribution given by Eq. (6-32). An appropriate transform for that function is

$$G(p_n) = \sum_{n=0}^{\infty} \left(1 - \frac{\lambda}{\mu}\right)\left(\frac{\lambda z}{\mu}\right)^n = \frac{\mu - \lambda}{\mu - \lambda z} \tag{6-37}$$

Using the nesting arguments of Chapter 4, it therefore follows that the Laplace transform for the distribution of the last n terms in the summation of Eq. (6-34) is

$$\begin{aligned} \mathscr{L}\left(f\left(\sum_{i=1}^{n} t_i\right)\right) &= G(p_n)\Big|_{z = \mathscr{L}(f(t))} \\ &= \frac{\mu - \lambda}{\mu - \lambda[\mu/(\mu + s)]} = \frac{(\mu - \lambda)(\mu + s)}{\mu(\mu - \lambda + s)} \end{aligned} \tag{6-38}$$

The transform for the total wait is then the product of the above transform and the transform of the distribution for t_0. That is,

$$\begin{aligned} \mathscr{L}(f(t_w)) &= \mathscr{L}(f(t))\mathscr{L}\left(f\left(\sum_{i=1}^{n} t_i\right)\right) \\ &= \left(\frac{\mu}{\mu + s}\right)\frac{(\mu - \lambda)(\mu + s)}{\mu(\mu - \lambda + s)} = \frac{\mu - \lambda}{\mu - \lambda + s} \end{aligned} \tag{6-39}$$

This is easily recognized as the Laplace transform of an exponential distribution with a parameter of $(\mu - \lambda)$. The waiting or delay time for the (M/M/1) system is therefore distributed by

$$f(t_w) = (\mu - \lambda) \exp\left[-(\mu - \lambda)t\right] \tag{6-40}$$

Similar arguments can be used to develop the waiting-time distributions for other system configurations in which the steady-state distribution p_n is known.

The Imbedded Markov Chain

One of the most frequently violated assumptions in applying the simple birth–death queueing model is that of exponentially distributed service times. The Poisson arrival pattern seems to fit many physical systems but systems with a very large number of extremely small service times are more rare. In order to provide a higher degree of isomorphism with the real world, we now want to explore systems with general service time distributions which may be decidedly different from the exponential distribution. Our vehicle for discussion will be the (M/G/1) system. The technique used is that of imbedded Markov chains first introduced by Kendall [23, 24].

The general service-time distribution introduces new complications since knowledge of the number of customers in the system at time t is no longer sufficient to predict the number present at time $t + \Delta t$. Consider, for example, a uniform service time distribution in the interval [2, 5]. If we examine the system 1 minute after service starts on the customer currently occupying the system, the probability of that customer being discharged in the next second is zero. If we examine the system 4 minutes after service starts, the probability of discharge in the next second is the conditional probability of a service time falling between 4 minutes and 4 minutes plus 1 second, given that it exceeded 4 minutes. Predicting system behavior then requires knowing not only how many customers are present, but also how long the lead customer has been in service. This system is no longer a simple Markov process. No continuous distribution other than the exponential exhibits the convenient property of forgetfulness. What we require for the general service case is some way to make the system look as though it is Markovian when it really is not.

Fixed Cycle Service

In order to illustrate the basic modeling approach, consider a fixed cycle conveyor system in a paint booth for bicycle frames with a single pickup hook. That hook passes the loading station every c minutes. If any frames are waiting, one will be loaded onto the conveyor, but the conveyor continues to cycle with or without a loaded hook. Further, suppose that as the hook leaves the loading station, a camera is tripped which records the number of frames waiting to be loaded. The only times data are recorded are at the instants just after service has taken place. Nothing is known about line length during the middle of the c-minute conveyor cycle. To complete the scenario, suppose that frames arrive at the loading station in Poisson fashion at a rate of a per minute and that there is no limit to storage capacity at the station. The task is to develop a probability distribution for the number of frames waiting to be loaded.

By taking snapshots of the system at a discrete sequence of times, we have extracted a discrete parameter process from the continuous process in which N_t might represent the number of frames present at any arbitrary point in time. What we see instead is a sequence of values for queue length only at those instants immediately following service. That is,

$$\{N_{T_i}\} = \{N_{T_1}, N_{T_2}, \ldots\} \tag{6-41}$$

where N_{T_i} represents the number in the queue at the time the ith cycle is completed. We will refer to this discrete parameter process, which has been extracted from the continuous parameter process, as an *imbedded Markov chain.*

The number of arrivals that occur during one cycle of the conveyor system will be Poisson distributed. Let

$$f_m = \Pr\ (m \text{ frames arrive during one cycle})$$

Then

$$f_m = e^{-ac}(ac)^m/m! \tag{6-42}$$

The transition matrix for changes occurring between successive occasions of service can be developed as follows:

1. If no frames were waiting at the time of the last snapshot, $j > 0$ will be waiting at the time of the next snapshot if and only if $j + 1$ arrive during the cycle time c. The extra arrival is needed since one of the newly arrived units will be served before the snapshot takes place. The system may move from zero to zero if either no arrivals or one arrival takes place in time c.
2. If $i > 0$ frames were waiting at the time of the last snapshot, $j + i$ will be waiting at the time of the next snapshot if and only if $j + 1 - i$ arrive during the cycle time c.

The matrix that results from these arguments is

$$\mathbf{T} = \begin{array}{c|ccccc} 0 & f_0 + f_1 & f_2 & f_3 & f_4 & \cdots \\ 1 & f_0 & f_1 & f_2 & f_3 & \cdots \\ 2 & 0 & f_0 & f_1 & f_2 & \cdots \\ 3 & 0 & 0 & f_0 & f_1 & \cdots \\ \vdots & \vdots & & & & \end{array} \tag{6-43}$$

If the system had a waiting room capacity of k frames, the matrix would have $k + 1$ rows and columns, with the last entry in row i being the sum of probabilities for more than $k - i$ arrivals.

Following the usual techniques for finding the steady-state solution, we must find a vector **P** such that

$$\mathbf{P} = \mathbf{PT} \tag{6-44}$$

The system of equations resulting from this calculation is

$$\begin{aligned} p_0 &= p_0(f_0 + f_1) + p_1 f_0 \\ p_1 &= p_0 f_2 + p_1 f_1 + p_2 f_0 \\ p_2 &= p_0 f_3 + p_1 f_2 + p_2 f_1 + p_3 f_0 \\ &\vdots \\ p_n &= \sum_{i=0}^{n+1} p_i f_{n+1-i}, \qquad n \geq 1 \end{aligned} \tag{6-45}$$

To seek a solution by geometric transforms, let us multiply the nth equation by z^n and add. The resulting single equation is then

$$\sum_{n=0}^{\infty} p_n z^n = p_0 f_0 + \sum_{n=0}^{\infty} \sum_{i=0}^{n+1} p_i f_{n+1-i} z^n \tag{6-46}$$

Let us define the geometric transform for the state distribution as

$$G(p_n) = P(z) = \sum_{n=0}^{\infty} p_n z^n$$

and the transform for the arrival distribution as

$$F(z) = \sum_{n=0}^{\infty} f_n z^n$$

We can use the convolution property of geometric transforms to expedite the reduction of Eq. (6-46). Rewrite the second term on the right as

$$\sum_{n=0}^{\infty} \sum_{i=0}^{n+1} p_i f_{n+1-i} z^n = \frac{1}{z} \sum_{k=1}^{\infty} \sum_{i=0}^{k} p_i f_{k-i} z^k$$

The inner sum can be recognized as the convolution of the p and f sequences. That is,

$$\sum_{i=0}^{k} p_i f_{k-i} = p * f$$

the transform for which is $P(z)F(z)$. It therefore follows that this expression can be written as a function of the appropriate transforms:

$$\frac{1}{z}\sum_{k=1}^{\infty}\left(\sum_{i=0}^{k} p_i\, f_{k-i}\right) z^k = \frac{1}{z}[P(z)F(z) - p_0\, f_0] \tag{6-47}$$

Using this result in Eq. (6-46), we now find

$$P(z) = p_0\, f_0 + \frac{1}{z}[P(z)F(z) - p_0\, f_0]$$

from which

$$P(z) = \frac{(z-1)p_0\, f_0}{z - F(z)} \tag{6-48}$$

However, the f_m distribution is the familiar Poisson distribution, from which

$$f_m = (ac)^m e^{-ac}/m!, \qquad F(z) = e^{(z-1)ac}, \qquad \text{and} \qquad f_0 = e^{-ac}$$

Therefore

$$P(z) = \frac{(z-1)e^{-ac}p_0}{z - e^{(z-1)ac}} \tag{6-49}$$

The initial condition p_0 can be recovered by remembering that

$$P(z)\Big|_{z=1} = 1 \tag{6-50}$$

Direct substitution leads to the indeterminant form 0/0. However, by applying L'Hospital's rule to the right side, which states that

$$\lim_{z\to 1}\frac{f(z)}{g(z)} = \lim_{z\to 1}\frac{f'(z)}{g'(z)}$$

we arrive at

$$1 = \frac{e^{-ac}p_0}{1 - ac}$$

from which

$$p_0 = (1 - ac)e^{ac} \tag{6-51}$$

It is obvious that the product ac must be less than unity for this solution to have any meaning. Recall that a is the customer arrival rate and c is the mean time between services. Equivalently, $1/c$ is the mean service rate. Simply stated, this means that the arrival rate must be less than the service rate before a stationary distribution can exist.

Returning to Eq. (6-49), the final expression for the transform is

$$P(z) = \frac{(z-1)(1-ac)}{z - e^{(z-1)ac}} \tag{6-52}$$

The distribution for the number of units waiting at those instants immediately following service and its moments can now be recovered by appropriate manipulation of the transform. It should be noted that Eq. (6-52) can be used as the solution for any fixed cycle system with cycle time c and Poisson arrival rate a.

Random Service Times

With the analysis of the fixed cycle system firmly in mind, it takes very little additional effort to develop a transform for the (M/G/1) system. We still retain the concept of snapshots of the system at those instants immediately after service takes place, i.e., we create an imbedded Markov chain. The only difference in visualizing the system with random service times, as opposed to a fixed service cycle, is that the time between transitions is now a random variable. Arrivals are assumed to be Poisson at a rate λ.

With random service it is now necessary to modify Eq. (6-42) to obtain the probability of m arrivals during a period of random length as opposed to the previous fixed transition time. Treating the number of arrivals in time t as a conditional event, and letting the general service time distribution be $g(t)$, we have

$$f_m = \int_0^\infty f(m, t)\, dt = \int_0^\infty f(m \mid t) g(t)\, dt \tag{6-53}$$

But the conditional distribution $f(m \mid t)$ is the Poisson distribution with mean λt. Therefore the probability of m arrivals between transitions is given by

$$f_m = \int_0^\infty \frac{e^{-\lambda t}(\lambda t)^m}{m!} g(t)\, dt \tag{6-54}$$

By keying transitions to the completion of service, as opposed to the completion of a fixed cycle time, we are led to one slight difference in the form of the transition matrix of Eq. (6-43). The first row of that matrix must be modified as follows:

$$\mathbf{T} = \begin{bmatrix} f_0 & f_1 & f_2 & \cdots \\ f_0 & f_1 & f_2 & \cdots \\ 0 & f_0 & f_1 & \cdots \\ \vdots & & & \end{bmatrix} \tag{6-55}$$

The first row is identical to the second because the arrival count cannot begin unless a unit is actively engaged in service. If the system is empty at a given service completion epoch, the next time the system will be examined is the next time a unit is discharged. But such a discharge cannot take place until a new unit arrives and then is served. In some sense we could think of the system as having an arrival counting clock which stops when there are no customers present. A move from zero to j between successive service epochs then requires that j customers arrive during the service of one customer, just as does the move from one to j. The only difference is that by starting in state zero we have extended the time during which system status is unknown by the time it takes to get the first new customer into our empty system. From the point of view of counts taken at completion of service, this is immaterial.

Let us now perform the same sequence of calculations illustrated by Eqs. (6-44)–(6-52). The steady-state balance equations are given by

$$p_n = p_0 f_n + \sum_{i=1}^{n+1} p_i f_{n+1-i} \tag{6-56}$$

for all n. Multiplying the nth equation by z^n and summing, we obtain

$$\sum_{n=0}^{\infty} p_n z^n = p_0 \sum_{n=0}^{\infty} f_n z^n + \sum_{n=0}^{\infty} \left[\sum_{i=1}^{n+1} p_i f_{n+1-i} \right] z^n$$

from which

$$P(z) = \frac{p_0(z-1)F(z)}{z - F(z)} \tag{6-57}$$

The geometric transform $F(z)$ can be expressed in closed form as a function of the Laplace transform of the service-time distribution as follows:

$$\begin{aligned} F(z) &= \sum_{m=0}^{\infty} f_m z^m = \sum_{m=0}^{\infty} \left[\int_0^{\infty} \frac{e^{-\lambda t}(\lambda t)^m}{m!} g(t)\, dt \right] z^m \\ &= \int_0^{\infty} \left[\sum_{m=0}^{\infty} \frac{e^{-\lambda t}(\lambda t z)^m}{m!} \right] g(t)\, dt \end{aligned} \tag{6-58}$$

But the term in brackets is simply the geometric transform for the Poisson distribution, or

$$G(f(m \mid t)) = e^{-\lambda t(1-z)}$$

Therefore we can rewrite Eq. (6-58) as

$$F(z) = \int_0^{\infty} e^{-\lambda t(1-z)} g(t)\, dt \tag{6-59}$$

But this expression looks like the Laplace transform of the service distribution with the transform argument s replaced by $\lambda(1 - z)$. It then follows that $F(z)$ can be written as

$$F(z) = \mathscr{L}(g(t))\Big|_{s=\lambda(1-z)} = L[\lambda(1 - z)] \tag{6-60}$$

where $L(s)$ represents the Laplace transform of the service distribution.

Returning to the line length distribution, we have

$$P(z) = \frac{p_0(z - 1)L[\lambda(1 - z)]}{z - L[\lambda(1 - z)]} \tag{6-61}$$

The initial condition p_0 is evaluated as before using

$$P(z = 1) = 1$$

from which, after applying L'Hospital's rule, we obtain

$$1 = p_0 \lim_{z \to 1} \frac{L[\lambda(1 - z)] - \lambda(z - 1)L'[\lambda(1 - z)]}{1 + \lambda L'[\lambda(1 - z)]} = p_0 \frac{L(0)}{1 + \lambda L'(0)} \tag{6-62}$$

But the transform of the service distribution evaluated at $s = 0$ is unity and its derivative is the mean value. It then follows that

$$p_0 = 1 - \lambda E(t) \tag{6-63}$$

Finally the transform for the line length distribution becomes

$$P(z) = \frac{(1 - \lambda E(t))(z - 1)L[\lambda(1 - z)]}{z - L[\lambda(1 - z)]} \tag{6-64}$$

This expression can now be used to develop the moments of the distribution by our usual transform techniques. Details of these calculations are left as an exercise for the reader.

Summary

The purpose of this chapter has been to demonstrate the usefulness of transform techniques in the context of a particular area of stochastic processes. The study of queues per se has been incidental to our purpose. Queueing theory is a very rich field for developing practical engineering tools. The demonstration that transforms provide convenient techniques for their development has been the intent of our study. For more details about such models the reader is urged to consult the references at the end of the book.

Problems

1. Consider the birth–death model of Eq. (6-30).

(a) List the assumptions necessary to use this model to analyze a real system.

(b) Illustrate the use of this model in analyzing a machine repair facility which must service 5 machines with a crew of 2 mechanics. State all assumptions used in constructing your model.

(c) Use numbers of your own choice to calculate the probability of at least one idle mechanic.

2. Consider each of the following as they relate to queueing models:

(a) Distinguish between stationary and steady state.

(b) What role does the Markov assumption play in the development of queueing models? Explain the difference in state variable definition between a system with negative exponential service and one with a general service distribution.

(c) List three general techniques for analyzing queueing systems in which the negative exponential service-time assumption is not valid. Explain the advantages and shortcomings of each.

3. Consider a maintenance facility with 2 first-line mechanics and 2 helpers who must maintain 4 pieces of equipment. The mean operating time between failures for each piece of equipment is 20 hours. When a single machine is inoperative all 4 maintenance men work on it and, on the average, can restore it to service in 4 hours. When 2 or more machines are inoperative, the group splits into 2 crews, each consisting of 1 mechanic and 1 helper. The average service time of each crew on a machine is 6 hours. Each hour that a machine is inoperative costs the company $100 in lost production. State all of your assumptions in answering the following questions:

(a) What portion of the time will all equipment be operative?

(b) What is the average number of machines requiring maintenance at any one time?

(c) What are the mean and variance for the amount of money lost per machine breakdown?

4. Develop numerical values for the probability distribution for number of busy telephone lines in an exchange containing 4 lines with no facilities for holding subscribers who require a line but cannot be supplied with one. The probability that a call will last less than t minutes is given by $1 - e^{-t/4}$. The time between arriving calls is distributed by $0.125e^{-t/8}$.

5. A marriage counselor has established the policy that he will accept only couples for consultation even though he interviews only one partner at a time. The probability that the time between arrivals of couples is less than t can be approximated by $1 - e^{-\lambda t}$. Furthermore, his service time for each person is distributed by $\mu e^{-\mu t}$.

(a) Develop an appropriate set of linear difference equations that will yield the equilibrium probabilities of finding n persons in the system. State all assumptions.

(b) It has been suggested that the counselor should consider counseling single persons as well as couples. If he does, it is estimated that when an arrival event occurs it will be a couple with probability q, otherwise it will be a single person. A single-server model with exponential interarrival times and exponential service can be used to model this new situation. Exhibit the linear difference equations that will capture the new system.

6. Modify the basic Poisson process such that each arrival event signifies the arrival of two customers and the time between events is negative exponentially distributed.

(a) Form a Markov transition matrix.

(b) Develop an appropriate set of differential–difference equations.

(c) Find the probability distribution for number of arrivals by time t assuming zero in the system to start.

7. Consider a maintenance facility which handles radio maintenance for a small feeder airline. The airline has a total of 4 radios to be maintained. The mean time between failures on any radio is estimated at 5 days. There are 2 mechanics available for radio maintenance. Service time is a random variable with an effective rate of 0.25 per day for each mechanic.

(a) Develop a model to describe the behavior of this maintenance system. State all assumptions.

(b) Find the expected number of mechanic days per year devoted to radio maintenance.

8. The University Registrar has a policy of adjusting service rate to customer demand. He uses his clerks for other tasks when they are not employed at the service windows. Consequently he requires only his newest clerk to be on duty until there is an accumulation of x customers in the system. At that time he will call a second clerk, who works at a speed of 1.2 times as fast as the new clerk. Both are on duty as long as x or more customers are present. However, when the population falls below x, the second clerk returns to standby status. When the population exceeds y (where $y > x$) the registrar joins the service mechanism, to give a total of 3 servers. He can turn out 3 customers while clerk number two processes 2 customers. When the level returns to y, the registrar leaves the serving queue and operates solely as a supervisor. The capacity of the waiting room is z customers. When z are present all others are denied services. Customers always arrive in pairs even though they are served singly. Time between customer pair arrivals is m minutes. The time to serve a single customer by a new clerk is n minutes.

(a) Develop a model to describe the number of customers present. State all assumptions.

(b) What conditions are necessary to have a stable system?

(c) Find the probability of no customers queued when $x = 2$, $y = 3$, $z = 4$, $m = 10$ minutes, and $n = 20$ minutes. How many customers can the registrar expect to turn away (deny service) in an 8-hour shift?

9. In a multichannel system with S servers a customer joins a single queue only if all service channels are occupied. Consider the special case of an (M/M/S) system.

(a) What portion of the arriving customers must join the queue?

(b) For those customers who must join the queue, find the probability that more than t minutes will be required before they reach a service facility.

(c) Find the probability that the total time from arrival to discharge for a customer will exceed t.

10. Use Eq. (6-64) to prove that the expected line length for an (M/G/1) queue is

$$E(n) = \lambda\tilde{t} + \frac{(\lambda\tilde{t})^2[1 + (\sigma^2/\tilde{t}^2)]}{2(1 - \lambda\tilde{t})}$$

where $\tilde{t}$ and σ^2 are the moments of the service distribution.

11. Consider the stream of customers departing from an (M/M/1) system.

(a) Prove that the time between customers leaving the system has the same distribution as the time between customer arrivals.

(b) Comment on the usefulness of this result in analyzing networks of queues. What assumptions must be made about service facilities hooked in series?

12. Suppose that a service system receives customers from two distinct sources. Customers from the first source arrive in Poisson fashion at a rate λ_1 and from the second source at a rate λ_2. Find the total arrival distribution for the service facility.

13. Suppose that customers departing from an (M/M/1) system are split in random fashion with a portion f_i of the total flow going to each of several new service facilities. Prove that the arrival process at each of the new facilities will be Poisson with rates $f_i\lambda$.

14. A used bicycle dealer operates a repair shop 8 hours per day to refurbish bicycles for resale. Bicycles are brought to his repair shop in Poisson fashion at a rate of 24 per day. The first operation is an inspection, which follows the exponential distribution with a mean of 15 minutes per bicycle.

After inspection 30% of the units require tire replacement, 40% require handlebar finishing, 20% require saddle repair, and 10% require new bearings. Furthermore, one-half of the units with tires replaced are also found at that time to need new bearings. The tire, handlebar, and bearing repair stations all operate in exponential fashion with rates of 4, 6, and 3 per hour, respectively. After completing all required repairs, every bicycle is painted. Painting time is uniformly distributed in the interval of [12, 18] minutes.

(a) Calculate the expected number of bicycles at each station in the repair network.

(b) Calculate the expected time for a randomly arriving unit to pass completely through the repair shop.

Hint: Use the results of Problems 10–13 to assist in your solution.

15. The local transit system operates a bus route with 3-seat jitneys. They pass the union pickup every t minutes, at which time they load all passengers present up to the capacity of the jitney. The vehicle then proceeds downtown, where all passengers are discharged. The union stop has a capacity of 5 customers. Customer arrivals can be considered to be Poisson at rate λ.

(a) Develop a model to describe the behavior of this system.

(b) Explain how you would find the portion of trips that the jitney will have a full load if $t = 30$ minutes and $\lambda = 9$/hour. Do you expect to be able to find a stationary solution? Why or why not?

16. When a modern-day Hercules takes on a modern Hydra he may find the head replacement problem a bit more complex. The modern Hydra is self-reproducing once the fight starts such that each of its heads at time t may split in two (become two new heads) in time $(t, t + h)$ with probability $\lambda h + O(h)$ and remain unchanged with probability $(1 - \lambda h) + O(h)$ as $h \to 0$. Hercules wants to know how many heads he will be facing at time t, say $N(t)$, if he confronts a one-headed Hydra at time zero.

(a) With $P(N(t) = j) = P_j(t)$, write the differential difference equations that determine $P_j(t)$.

(b) Show that $N(t)$ has the probability generating function

$$P(z, t) = \frac{ze^{-\lambda t}}{1 - z + ze^{-\lambda t}}$$

(c) Find $E(N(t))$.

(d) Find $P_j(t)$.

(e) Find $P(z, t)$ if Hydra begins the fight with $n_0 > 1$ heads.

17. A training clinic for medical doctors has a normal complement of 3 interns. An average of 4 patients per hour seek services at the clinic. When a patient arrives he must be examined by one or more of the interns. When only 1 patient is present 1 intern will see that patient with probability 0.4, 2 with probability 0.5, and 3 with probability 0.1. They examine in series and each follows a service distribution given by $2e^{-2t}$. When 2 patients are present the first one in the examining room will see 1 doctor with probability 0.7 and 2 with probability 0.3. When 3 or more patients are present all doctors know that they will be able to examine someone if they rotate in the examining room. Therefore the lead patient under those conditions will see only 1 doctor. Since there is only one examining room, no new patients can be admitted to the examining room until all doctors who are scheduled to see the current patient complete their examination. Waiting room space is limited to a total of 5 patients, not including the one in the examination room.

(a) Develop an appropriate model to describe the probability distribution for number of patients at the clinic. State all assumptions.

(b) Explain in detail how you would obtain a solution for your model and accompanying performance measures.

18. Consider a bulk service queueing problem in which batches of waiting units are served. If the line is longer than m units, the batch size is m. If the line contains m or fewer units at an occasion of service, all waiting units are served. Customers arrive in Poisson fashion at a rate λ. Service is instantaneous but only available at service instants, the intervals between which are independently distributed by $b(t)$. The random variable q_n represents the number of customers in the system just before the nth service instant. The random variable x_n represents the number of arrivals between the nth and $(n + 1)$th service instants.

(a) Develop a recursion formula relating q_{n+1} and q_n.

(b) Use your recursion formula and your knowledge of transforms to show that

$$Q(z) = \frac{\sum_{i=0}^{m-1} P_i(z^m - z^i)}{z^m(B(\lambda - \lambda z)^{-1}) - 1}$$

where

$$P_i = \text{probability that} \quad q_n = i$$

$$Q(z) = \sum_{i=0}^{\infty} z^i P_i \qquad \text{and} \qquad B(s) = \mathscr{L}(b(t))$$

19. A methods analyst has rigged a camera to shoot one frame of film at the instant an engine leaves the test stand. The film records the number of engines waiting to be tested. The test facility can accommodate no more than 2 engines in addition to the one which may be in the process of being tested. All engines that arrive when the system is full are denied service and must be shipped to a competitor for testing. As long as the test facility remains in continuous operation, the time to test an engine x is distributed by

$$f(x) = 2e^{-2x}$$

where x is measured in days. Each time that the queue becomes empty the test stand is shut down. When it is next started to test an engine, there is a fixed warmup period of $\frac{1}{2}$ day, after which the usual random production time takes place. The probability of less than t days between arrivals of engines is given by

$$1 - e^{-3t}$$

(a) Find the limiting probability distribution for number of engines that will be recorded on each frame.

(b) Find the expected number of engines shipped to a competitor in a 24-day period.

20. Consider a single data link which can occupy only two states. Either it is busy or it is inactive. Transmission attempts occur in Poisson fashion at a rate λ. The time it takes to transmit a single data set is exponentially distributed.

(a) Determine the set of differential–difference equations that describes the behavior of the system.

(b) Find the probability that the system will be busy t minutes from now if it is currently busy.

(c) Find the long-run utilization of the data link.

CHAPTER 7

LINEAR SYSTEMS ANALYSIS

The purpose of this chapter is to introduce to the reader a slightly different application for transform techniques in models which may be encountered in analyzing man–machine systems. The method of presentation will be to describe the development of transfer functions for deterministic systems in a manner similar to that found in early literature on servomechanisms. Although the main thrust of this material is not probability modeling, after building that foundation similarities with certain basic stochastic process models will be noted. The intent is to provide the reader with a useful method for organizing system structures into model representations and to further his skills in manipulating the various transforms discussed earlier. Readers who require more depth in these topics should consult the excellent presentation by Howard [21].

Linear Systems

In simplest terms we will treat a "system" as a mathematical operator which operates on one signal or function to produce another. The physical counterpart might be an automobile driver, an accounting system, or an automatic conflict resolution program in an air traffic control radar system.

In the case of an automobile driver, the human system may be stimulated by receiving a series of light signals designed to convey information to him. He responds by providing an output function in the form of accelerator pedal deflection, for example. What we seek is some way of representing this connection between input and output by a mathematical operator, which we can then use to predict a driver's response to new stimuli or input signals. The driver is treated as a "black box" with observable inputs and outputs but unknown inner workings.

In the case of an accounting system, the input signal may be the string of customer demands placed upon an inventory system. The output signal may be purchase orders from the holder of the inventory to his vendors for resupply. The connection between these two signals is the system or operator that we seek to model.

In the case of automatic radar tracking in an air traffic control system, the input signal is the recorded position of an aircraft on successive sweeps of the radar antenna. The output is a predicted position of that aircraft at some time in the future. That predicted position will indicate whether or not evasive action may be required to prevent collision with other known traffic. The system or operator is the device that functionally relates the input and output signals.

Linear systems are only a small subset of all possible system configurations one might consider. For a system to be linear, signals must be additive. Consider the case of an inventory control system for transmitter crystals in an aircraft radio repair shop. Crystals may be supplied to external customers who arrive in a stream described by the function $f_1(n)$. In addition, crystals may be required to repair the company's own radios, which fail according to the time function $f_2(n)$. Suppose that it has been observed that vendor purchase orders are issued in a stream $g_1(n)$ when only external customers are served and in a stream $g_2(n)$ when only their own radios are repaired. Now suppose that business increases such that the normal stream of external demands is multiplied by a and internal demands by b. If the inventory control system is linear, the new stream of vendor purchase orders will be

$$g(n) = ag_1(n) + bg_2(n) \tag{7-1}$$

It is obvious that many real-world systems are nonlinear, especially when considered throughout a large range of input signals. The human operator, for example, has been proven many times over to be an annoyingly nonlinear system. Production–inventory control systems may be nonlinear at extreme ends of their operating range. Increase in production output, for example, may not be proportional to the number of man-hours added to a system already operating to near capacity, due to inefficiencies brought about by congestion and overtime operations. When faced with such nonlinearities, our response will be to attempt to model the system as though it is linear with strong caveats against applying the results outside a very narrowly defined range. If the system under study is violently nonlinear even over limited ranges of input and output signals, the techniques of this chapter are not applicable.

Superposition

The defining characteristic of the linear systems which we have been discussing is given the name superposition. More formally stated, suppose a system is given an input signal f_1 and the resulting output g_1 is observed. Then it is given a signal f_2 and the resulting output g_2 is observed. A new signal is then formed by multiplying each of the input signals by a constant and adding. If the resulting output from the system is equal to the sum of the original separate outputs each multiplied by the corresponding input constant, the system is said to be linear.

The property of superposition is equally applicable for both continuous and discrete systems. In the context of continuous systems it is convenient to think of a linear system as one which can be described by a linear differential equation. A differential equation is linear when the dependent variable and its derivatives appear only to the first degree. For example, the equation

$$c_2 \frac{d^2 g(x)}{dx^2} + c_1 \frac{dg(x)}{dx} + c_0 g(x) = f(x) \tag{7-2}$$

is a second-order, linear differential equation with constant coefficients. In our use of systems analysis we will think of $f(x)$ as the driving or input function and $g(x)$ as the response or output function.

For example, suppose that $g_1(x)$ is the solution to Eq. (7-2) when $f_1(x)$ is the driving function and that $g_2(x)$ is the solution when $f_2(x)$ is the driving function. That is,

$$c_2 \frac{d^2 g_1(x)}{dx^2} + c_1 \frac{dg_1(x)}{dx} + c_0 g_1(x) = f_1(x) \tag{7-3}$$

and

$$c_2 \frac{d^2 g_2(x)}{dx^2} + c_1 \frac{dg_2(x)}{dx} + c_0 g_2(x) = f_2(x) \tag{7-4}$$

By adding Eqs. (7-3) and (7-4), we obtain

$$c_2 \frac{d^2}{dx^2}[g_1(x) + g_2(x)] + c_1 \frac{d}{dx}[g_1(x) + g_2(x)] + c_0[g_1(x) + g_2(x)]$$
$$= f_1(x) + f_2(x) \tag{7-5}$$

It is apparent that the system, when subjected to a new input $f_1(x) + f_2(x)$, will have a new response $g_1(x) + g_2(x)$. This is simply the sum of the responses to the inputs considered separately, which is in keeping with our earlier definition of superposition.

Time Invariance

A second property of those linear systems that are of concern to us is that of time invariance. Suppose that a system responds to an input function $f(n)$ applied at time zero with an output function $g(n)$. If the input function is shifted such that the origin is now at $n = k$, the new output signal, which is identical in form to the old one, should be shifted a like amount. In other words, if $f(n)$ leads to $g(n)$, then $f(n - k)$ will lead to $g(n - k)$, where the system is at rest prior to $n = k$. That is the property of time invariance.

In the context of a production–inventory system, suppose that a sale of 5 units on the 1st day of the month results in a work-release order to the shop for 6 replacement units on the 10th of the month. If the system is time invariant, a sale of 5 units on the 15th of the month should trigger a work-release order for 6 units on the 24th of the month.

Impulse Response Functions

The response of a system initially at rest to a unit impulse input is a very important function for our later analysis. In discrete systems the unit impulse is simply a spike of magnitude one at the origin. Such a case might be a sales demand for a single unit in an inventory system. In continuous systems the unit impulse can be viewed as the Dirac delta function applied at the origin. Recall that the properties of this function are

$$\delta(t - a) = 0, \quad t \neq a,$$

and

$$\int_0^\infty \delta(t - a)\, dt = 1$$

This is obviously a mathematical convenience. It is at best an approximation to the physical realities of subjecting a system to a pulse of large magnitude over a very short duration, as in studying the response of spring–mass systems to pulse loads.

The output of a system which has been driven by a unit impulse input, be it discrete or continuous, will be referred to as the impulse response function. We will reserve the symbols $h(n)$ for discrete systems and $h(t)$ for continuous systems to represent the impulse response function.

Convolutions in Linear Systems

The primary reasons that we are interested in studying linear systems are that they can be totally described by their impulse response functions and that one can easily combine simple system elements into larger, more complex systems. The key to these observations is once more a thorough understanding of the convolution operation.

Discrete Systems

Consider the properties of superposition and time invariance as related to the impulse response of a discrete linear system. Let us designate the discrete impulse at the origin as $I(n)$ and the corresponding impulse response as $h(n)$. If we were to shift the input pulse k units to the right, the new response would be $h(n-k)$ due to the time invariance of that system. Furthermore, if we were to multiply the input impulse function by a constant, say $f(k)$, the output would be multiplied by the same constant, say

$$g(n) = f(k)h(n-k) \tag{7-6}$$

Now suppose that we consider some general driving function $f(n)$. The value of that function for a particular point $n = k$ can be written as

$$d(k) = f(k)I(n-k) \tag{7-7}$$

If our system is assumed to have been subjected to a series of such pulses beginning at $k = 0$ and continuing up to time n, the total input can be represented as

$$d(n) = \sum_{k=0}^{n} f(k)I(n-k) \tag{7-8}$$

Because of the superposition and time invariance properties of linear systems, the corresponding output is seen to be

$$g(n) = \sum_{k=0}^{n} f(k)h(n-k) \tag{7-9}$$

Equation (7-9) is readily identified as the convolution sum. This important result says that we should be able to find the output signal from a linear system, when subjected to an arbitrary input function $f(n)$, by convolving the input function with the impulse response function for that system. The implication is that if we know the impulse response function for a linear system, we know all there is to know about the input/output characteristics of that system.

Example Consider an econometric model of research expenditures for a toy manufacturer. Suppose that it has been estimated that \$1 spent on research today will lead to saleable products which will contribute increased income of \$2 two years from now, \$3 in the third year, \$1 in the fourth year, and zero thereafter. The implied impulse response function is $h(n)$, where

n	$h(n)$
0	0
1	0
2	2
3	3
4	1
≥ 5	0

Now suppose that the company plans to spend \$10,000, \$15,000, \$20,000, and \$25,000 dollars on research over the next 4 years. The problem is to determine the income expected to be generated by this research program. The new input function is $f(n)$, where

n	$f(n)$
0	0
1	10
2	15
3	20
4	25
≥ 5	0

By applying Eq. (7-9) we are led to the expected income function $g(n)$ in thousands of dollars as follows:

n	$g(n) = \sum_{k=0}^{n} f(k)h(n-k)$
0	0
1	0
2	0
3	$(10)(2) = 20$
4	$(10)(3) + (15)(2) = 60$
5	$(10)(1) + (15)(3) + (20)(2) = 95$
6	$(15)(1) + (20)(3) + (25)(2) = 125$
7	$(20)(1) + (25)(3) = 95$
8	$(25)(1) = 25$
≥ 9	0

Continuous Systems

The continuous analog of the system described above is one in which the driving function is considered to be a series of pulses of magnitude $f(s)\,\Delta s$. When integrated, this small rectangle is approximated by a delta function $\delta(t - s)$ multiplied by the constant $f(s)\,\Delta s$. The output corresponding to a single such pulse at $t = s$ is the corresponding impulse response function multiplied by $f(s)\,\Delta s$, i.e., $h(t - s)f(s)\,\Delta s$. By summing a series of such inputs to approximate the driving function $f(t)$, the corresponding output, expressed in the limit as Δs approaches zero, becomes

$$g(t) = \int_0^t f(s)h(t - s)\,ds \tag{7-10}$$

This integral is the convolution of the driving function $f(t)$ and the impulse response function $h(t)$. It plays the same role for continuous systems that Eq. (7-9) plays for discrete systems.

Transfer Functions

Because of the presence of the convolution operation in the analysis of linear systems, it should come as no surprise that the various transform techniques introduced in Chapter 2 may be useful. For purposes of this chapter we will employ the geometric transform when operating with discrete systems and the Laplace transform when operating with continuous systems. Let us replace the convolution sum of Eq. (7-9) and the convolution integral of Eq. (7-10) with the general convolution operator $(*)$, where it is understood to represent the sum when the functions are discrete and the integral when they are continuous. Then

$$g = f * h \tag{7-11}$$

By taking transforms (either geometric or Laplace as the case may be) of both sides of this equation, we are led by the convolution property of transforms to

$$T(g) = T(f)T(h) \tag{7-12}$$

The transform of the output from a linear system can be expressed as the product of the transform of the input and the transform of the impulse response function.

By rearranging terms in Eq. (7-12), we are led to another important conclusion:

$$T(h) = \frac{T(g)}{T(f)} \tag{7-13}$$

This says that the transform of the impulse response function can be obtained by taking the ratio of the transform of the output to the transform of the input that generated that output. When written this way for a system that is initially at rest, the ratio is referred to as the *transfer function* for the system:

$$\text{transfer function} = \frac{\text{transform out}}{\text{transform in}} \tag{7-14}$$

As we have just demonstrated, the transfer function can be obtained by taking the transform of a known impulse response function or by taking the ratio of output to input transforms, using any convenient driving function, when the impulse response is not available. Since the function $T(h)$ in Eq. (7-12) is equivalent to the transfer function, it is obvious that the response transform for any new driving function can be obtained by simple multiplication:

$$\text{transform out} = (\text{transform in})(\text{transfer function}) \tag{7-15}$$

This calculation is the basis for all of the uses of transforms in linear systems analysis which follow.

Example Consider a sales forecasting model that uses simple exponential smoothing to develop estimates for future sales from the time series of past demand. The smoothed function at time n, $s(n)$, is obtained from past observed data $f(n)$ through the recursive relationship

$$s(n) = s(n-1) + \alpha[f(n) - s(n-1)] \tag{7-16}$$

where α is a smoothing constant falling in the interval [0, 1].

The transfer function for this forecasting system can be developed by operating on the difference equation with geometric transforms. Taking transforms of both sides, we obtain

$$G(s(n)) = G(s(n-1)) + \alpha G(f(n)) - \alpha G(s(n-1)) \tag{7-17}$$

By the shifting properties of geometric transforms, we know that

$$G(s(n-1)) = zG(s(n))$$

Therefore we can rewrite Eq. (7-17) as

$$G(s(n)) = zG(s(n)) + \alpha G(f(n)) - \alpha z G(s(n))$$

Solving for the ratio $G(s(n))/G(f(n))$, we see that the transfer function for this system is

$$T(h) = \frac{G(s(n))}{G(f(n))} = \frac{\alpha}{1 - (1-\alpha)z} \tag{7-18}$$

Now suppose that this system is being used in a sales environment in which sales are increasing in linear fashion,

$$f(n) = 10 + 2n$$

and a smoothing constant of $\alpha = 0.3$ is in use. The transform of this sales or input signal is

$$G(f(n)) = \frac{10}{1 - z} + \frac{2z}{(1 - z)^2}$$

Since the transform of the output is obtained by multiplying the transform of the input by the transfer function, we have

$$G(s(n)) = \frac{10(0.3)}{(1 - z)(1 - 0.7z)} + \frac{2(0.3)z}{(1 - z)^2(1 - 0.7z)}$$

Inverting this transform, we obtain the smoothed function

$$s(n) = 10 + 2n - \frac{0.7}{0.3}2 - 10(0.7)^{n+1} + \frac{2}{0.3}(0.7)^{n+1} \tag{7-19}$$

The last two terms in Eq. (7-19) will go to zero for large n but the constant (-4.67) remains to distinguish the forecasted value from the sales input. Simple exponential smoothing, in this case involving input with a trend, will always lag the actual time series by that amount.

Combined Systems

Because of the convolution properties of linear systems, it is now possible for us to analyze interconnected simple systems by the use of transforms. The basic configurations we will examine are series, parallel, and feedback loop connections. Very complex systems can be analyzed by breaking them into many simple independent systems whose interactions can be characterized by a number of such connections.

Consider a production–inventory control system, for example. Sales data generated at the retail level for many items are fed into the retail accounting system, which uses the data to update records, prepare forecasts, and issue replenishment orders to the factory. Many retail outlets may be hooked up in parallel to the factory warehouse. The factory warehouse in turn analyzes the retail data, issues requests for production, and resupplies the retail outlets. The retail outlets, factory warehouses, and production facility are hooked up in series. When production orders are released, that information goes via a feedback loop to the warehouse. When retail shipments are made, another feedback loop involving the retail accounting system is used.

One thing to bear in mind as we analyze these structures is that we are talking about information flow, not the flow of physical goods. An electrical circuit analogy is the difference between current and voltage. When the output from subsystem A is fed to subsystems B and C, both B and C see the same signal. The signal is *not* split with a portion going to each subsequent subsystem.

Block Diagrams

The easiest way to visualize the structure of compound systems is by block diagrams. Each subsystem is represented by a block with the arrow entering the block being the input signal and the arrow exiting being the output signal. The block itself is the mathematical operator that operates on the input signal to obtain the output.

We could draw such diagrams when the signals are time functions and the system is characterized by its impulse response function. However, it will be more convenient for our purposes to work in the transform domain. In our block diagrams, then, the functions associated with the arrows will be the transforms of the indicated signals and the system operator will be the transform of the impulse response function or the transfer function. Either geometric or Laplace transforms will be used, depending upon whether the system is discrete or continuous. When neither is specified, the general operator T_j will be used. The subscript h refers to the transfer function, g to the output signal, and f to the input signal. Figure 7-1 illustrates the standard form for our diagrams.

Because of the convolution properties of transforms, the output signal on the standard diagram can be written as the product of the input transform and the transfer function:

$$T_g = T_f T_n \tag{7-20}$$

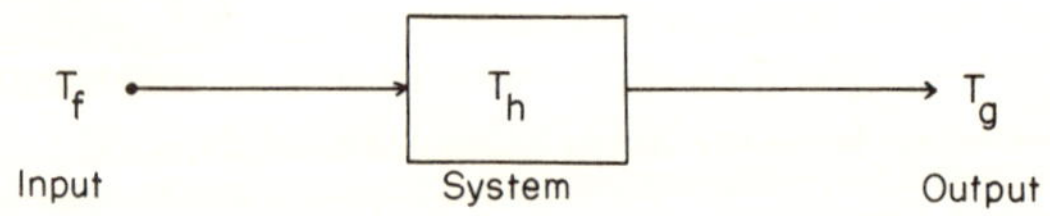

Figure 7-1 Basic block diagram.

Additivity

The first observation to make about interconnected systems is that multiple inputs to a single node or system are additive. If we have a system with k separate signals entering a common node, the total signal is

$$T_g = T_{g_1} + T_{g_2} + \cdots + T_{g_k} \tag{7-21}$$

In the context of our production–inventory control system this means that the total demand signal perceived by the factory warehouse is equal to the sum of the signals coming from the separate retail outlets. Graphically, this connection is shown by Figure 7-2.

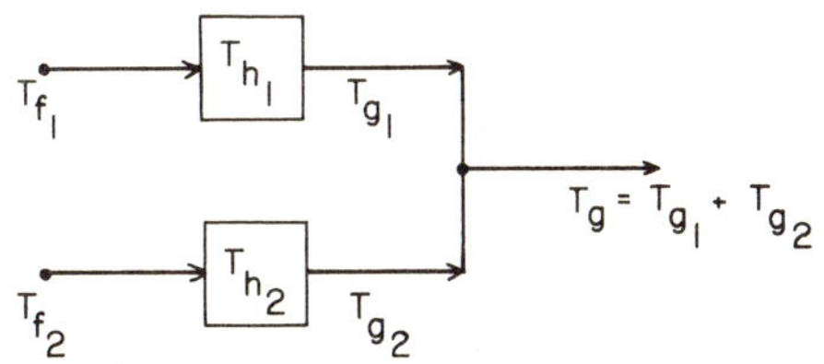

Figure 7-2 Additivity of signals.

Systems in Series

The object behind all of our manipulations of block diagrams is to reduce complex systems to a single block whose transfer function will replace the functions of all of the separate blocks constituting that system. To further that aim, let us first consider a simple two-element system in which the elements are connected in series. Such a connection might occur when sales signals are fed to the warehouse, which in turn feeds production requests to the factory. Figure 7-3 illustrates the series connection.

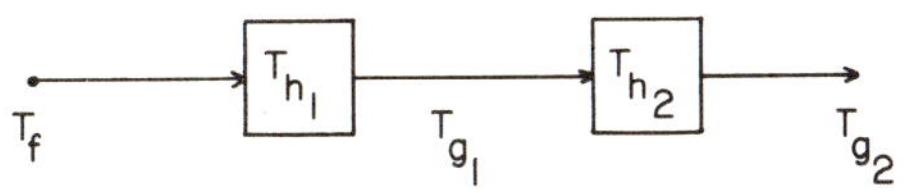

Figure 7-3 Elements in series.

Treating the first subsystem in isolation, the output signal from that element can be expressed as

$$T_{g1} = T_f T_{h1} \tag{7-22}$$

because of the convolution property of linear systems. Now turning our attention to the second subsystem, we have a single element with an input signal T_{g1}. It therefore follows that the output from the second element should be

$$T_{g2} = T_{g1} T_{h2} \tag{7-23}$$

But we have previously developed an expression for T_{g1} in Eq. (7-22). Combining terms, we find

$$T_{g2} = T_f T_{h1} T_{h2} \tag{7-24}$$

The form of Eq. (7-24) is identical to that of (7-12), where we first introduced the calculation of output signals by multiplication of transforms. The only difference is that we now have a product of two transforms in place of the single function of Eq. (7-12). It therefore follows that the transfer function for a single-element system that can be used to replace a series connection is given by

$$T_h = T_{h_1} T_{h_2} \tag{7-25}$$

The extension of this result to k elements in series is obviously the product

$$T_h = \prod_{i=1}^{k} T_{h_i} \tag{7-26}$$

Systems in Parallel

A second fundamental connection is one in which multiple elements are hooked up in parallel. In this configuration all the parallel elements receive the same input signal. Furthermore, all the outputs are also connected to a common node. Figure 7-4 illustrates the parallel connection.

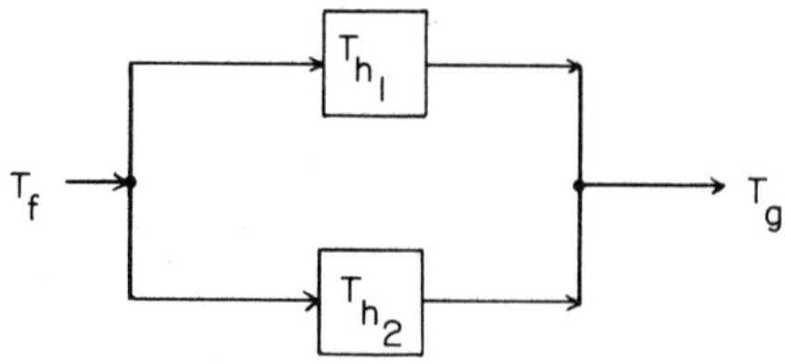

Figure 7-4 Elements in parallel.

We want to find a single transfer function which can be used to replace the combination of simple systems hooked up in parallel. From the diagram the input to each of the systems is the common function T_f. Both systems see the same signal. The output from system 1 is

$$T_{g1} = T_f T_{h_1} \tag{7-27}$$

and from system 2 is

$$T_{g2} = T_f T_{h_2} \tag{7-28}$$

The two output signals are then combined to yield

$$T_g = T_{g1} + T_{g2} \tag{7-29}$$

by the additivity property. Combining equations, we are led to

$$T_g = T_f T_{h_1} + T_f T_{h_2} = T_f(T_{h_1} + T_{h_2}) \tag{7-30}$$

From Eq. (7-30) it is obvious that the single element equivalent to the two-element parallel system has a transfer function of

$$T_h = T_{h_1} + T_{h_2} \tag{7-31}$$

The extension to k elements in parallel is straightforward and yields the sum

$$T_h = \sum_{i=1}^{k} T_{h_i} \tag{7-32}$$

Feedback Systems

The simplest feedback loop is one in which a third node exists to serve as both the input and output node of a system. The output signal for such a system becomes a function of itself. Figure 7-5 illustrates a simple feedback loop.

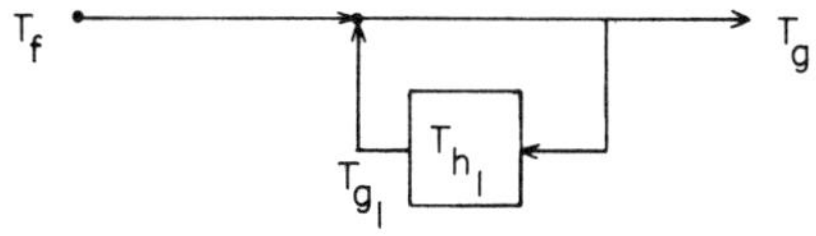

Figure 7-5 Simple feedback.

An automatic controller on a machine tool operates in such a fashion. The current diameter of pieces being machined is measured. The results of that measurement are then fed back through the controller, which makes adjustments to the tool setting. Those adjustments then influence the diameter of subsequent production items.

The output from the simple element in the loop is the product of its input function transform and its transfer function. But the input to that system is the total system output function T_g. It therefore follows that

$$T_{g_1} = T_g T_{h_1} \tag{7-33}$$

Suppose that we now treat this as a new signal, forgetting that its origin was the feedback element. Redrawing the diagram as in Figure 7-6, we simply have two signals being added at a common node. From the additivity

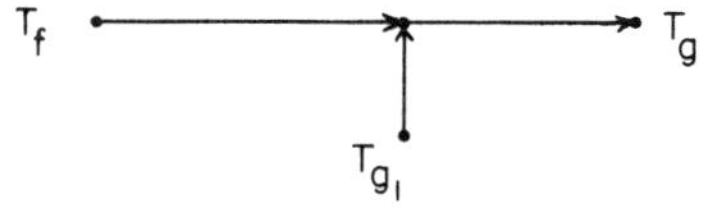

Figure 7-6 Feedback after modification.

property it follows that

$$T_g = T_f + T_{g1} = T_f + T_g T_{h_1} \tag{7-34}$$

Solving for T_g, we find that

$$T_g = T_f \frac{1}{1 - T_{h_1}} \tag{7-35}$$

Equation (7-35) is in the standard form of

transform out = (transform in)(transfer function)

It therefore seems reasonable to replace the feedback operation with a single simple element whose transfer function is

$$T_h = \frac{1}{1 - T_{h_1}} \tag{7-36}$$

A more complex feedback system exists when one element connects input to output and a second element connects output to input. We will refer to this as compound feedback. The diagram for such a system is presented in Figure 7-7.

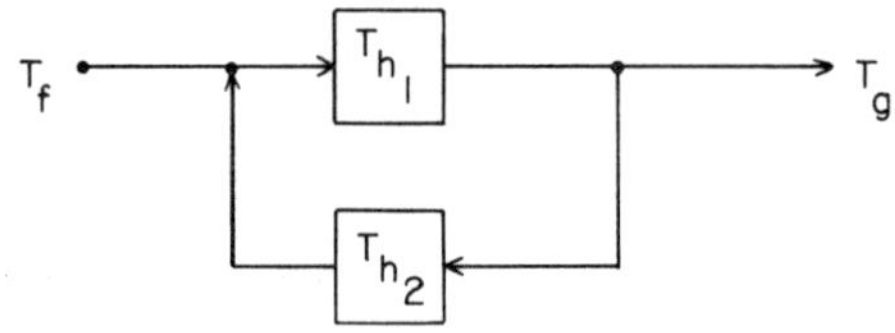

Figure 7-7 Compound feedback.

The key to reducing this system to a single simple element is to treat the output from the bottom element as a new input signal coming from some unidentified source just as we did in Figure 7-6. The input to the first element is then the sum of the outside signal T_f and the output from element 2. Therefore the output at element 1 can be written as

$$T_{g_1} = (T_f + T_{g2})T_{h_1} \tag{7-37}$$

The signal T_{g_1} also serves as the input to the second element. This means that the previously unidentified input to element 1, which is in reality the output from element 2, can be written as

$$T_{g_2} = T_{g_1} T_{h2} \tag{7-38}$$

Substituting back into Eq. (7-37) and recognizing that $T_{g_1} = T_g$, we obtain

$$T_g = (T_f + T_g T_{h_2})T_{h_1} \tag{7-39}$$

from which we find

$$T_g = T_f \frac{T_{h1}}{1 - T_{h1} T_{h2}} \tag{7-40}$$

When written in this manner it is obvious that the single simple element that can be used to replace the compound feedback loop has a transfer function of

$$T_h = \frac{T_{h1}}{1 - T_{h1} T_{h2}} \tag{7-41}$$

The results of Eqs. (7-21), (7-26), (7-32), (7-36), and (7-41) now provide us with sufficient flexibility to determine single transfer functions for a great many complex system structures.

Example Consider the block diagram of Figure 7-8. This system contains every type of connection discussed above. What we now seek is a single transfer function that will replace all of the simple systems shown.

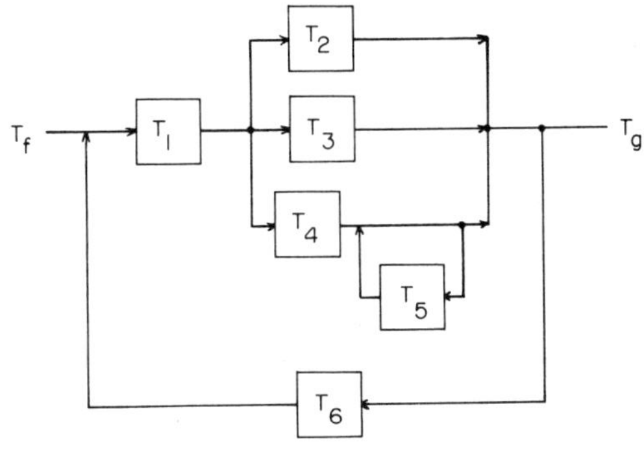

Figure 7-8 Complex system.

Let us begin the reduction of the system with the simple feedback loop containing T_5. That loop can be replaced by the simple element

$$T_7 = \frac{1}{1 - T_5}$$

But T_7 is hooked up in series with T_4, so that the entire branch can be replaced by

$$T_8 = T_4 T_7 = \frac{T_4}{1 - T_5}$$

This then leads to three elements in parallel, T_2, T_3, and the new element T_8.

These three can be replaced by the single element

$$T_9 = T_2 + T_3 + T_8 = T_2 + T_3 + \frac{T_4}{1 - T_5}$$

The new element T_9 is connected in series with T_1. Those two can be replaced by

$$T_{10} = T_1 T_9 = T_1\left(T_2 + T_3 + \frac{T_4}{1 - T_5}\right)$$

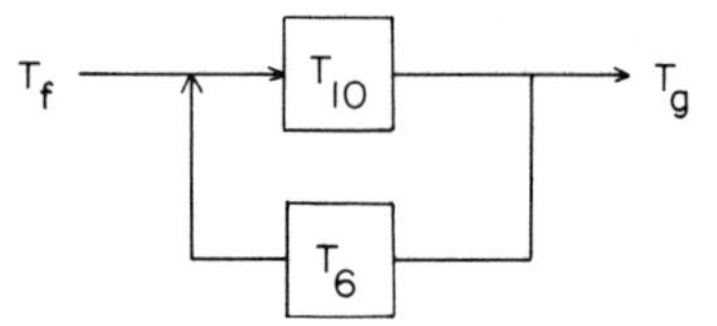

Figure 7-9 Partial reduction.

Let us pause now and examine our reductions so far. Figure 7-9 shows the new system, which can be seen to be a simple compound feedback loop involving T_{10} and T_6. Applying our final reduction to that loop, we find for the entire system

$$T = \frac{T_{10}}{1 - T_{10} T_6} = \frac{T_1\{T_2 + T_3 + [T_4/(1 - T_5)]\}}{1 - T_1\{T_2 + T_3 + [T_4/(1 - T_5)]\}T_6}$$

To study the characteristics of this complex system, we would now need to know only the transfer functions for each separate element of the system treated independently. This is a very powerful technique for both analysis and design of systems.

An Inventory Application

To place the use of our network reduction techniques in a real-world context, consider the following simple inventory ordering rule. An automobile dealer has a policy of attempting to maintain a 60-day supply of vehicles either on order or in stock. He uses simple exponential smoothing to forecast sales each month, with the smoothed value of the sales curve in month n being his estimate for sales in month $n + 1$. Orders are placed at the end of the month for enough vehicles to cover 60 days of estimated sales, less the number currently held in inventory. Vehicles produced in month n are delayed an additional 30 days before they reach the dealer's inventory. Inventory balance is calculated at the end of the month from the previous month's inventory, deliveries from the factory during the month, and sales for the month.

Let $s(n)$ be the sales during month $n, f(n)$ the smoothed value of the sales curve for month n, $i(n)$ the inventory on hand at the end of month n, and $p(n)$ the factory production for the dealer during month n. The inventory balance equation can be written as

$$i(n) = i(n-1) + p(n-1) - s(n) \tag{7-42}$$

The exponentially smoothed value for the sales curve, using a smoothing constant of a, is given by

$$f(n) = as(n) + (1-a)f(n-1) \tag{7-43}$$

Remembering that production for month n is a function of the smoothed sales curve and inventory level at the end of month $n-1$, the 60-day ordering policy yields

$$p(n) = 2f(n-1) - i(n-1) \tag{7-44}$$

In order to make the connection between the block diagram and the equations above, let us modify our past practice and represent each subsystem operator in terms of its impulse response function rather than going directly to the transform. The impulse response functions in the system are particularly easy to determine since most are simple delays. A system that delays a signal by one time period (1 month in this example) would have an impulse response function of $I(n-1)$, where

$$I(n-j) = \begin{cases} 1 & \text{for } n = j \\ 0 & \text{elsewhere} \end{cases}$$

A block diagram with the impulse response functions shown in each block is given in Figure 7-10. The reader is urged to trace the signal paths to

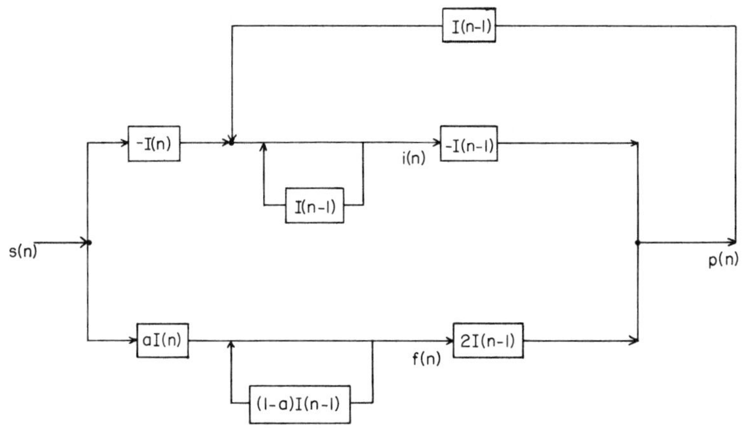

Figure 7-10 An inventory ordering rule.

verify the relationships among Eqs. (7-42)–(7-44). Conversion of the impulse response functions to transfer functions and reduction of the graph are left as exercises for the reader.

Multiple Input and Output Systems

Our next goal is to extend the concepts of transfer functions and network reduction to systems in which each subsystem may have multiple inputs and outputs. Now, rather than working with single functions, we will be using vectors and matrices to perform the necessary calculations.

As an example of such a multiple-node system, consider the cash flow through a local welfare agency. The agency may receive cash from public donations, labor union support, business contributions, and government grants, for a total of four separate input streams. The agency then operates on these streams and provides output in the form of aid to dependent children and aid to the handicapped for two output streams. Such a system operator is depicted in Figure 7-11.

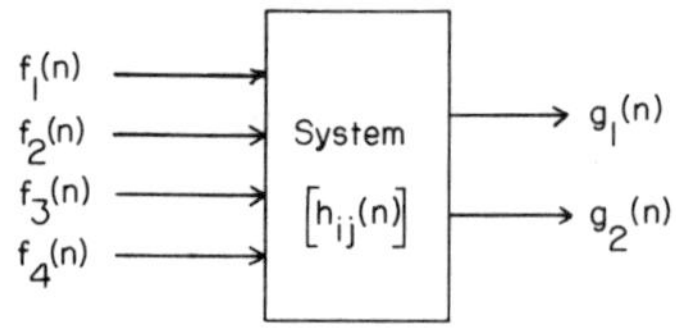

Figure 7-11 Multiple-node systems.

Matrix Representation

In the present context input and output streams to a subsystem will be represented by vectors. The impulse response function for the system is a matrix. The elements of the stream vectors are functions associated with the various system nodes. The elements of the impulse response matrix are impulse response functions for the nodes considered one pair at a time. The element $h_{ij}(n)$ represents the impulse response at node j when the system has been excited by a unit pulse at node i, assuming all other nodes are unexcited.

Applying these concepts to Figure 7-11, we can represent the input functions by the vector

$$\mathbf{f}(n) = (f_1(n), f_2(n), f_3(n), f_4(n))$$

the output by the vector

$$\mathbf{g}(n) = (g_1(n), g_2(n))$$

and the system operator by the matrix

$$\mathbf{H}(n) = \begin{pmatrix} h_{11}(n) & h_{12}(n) \\ h_{21}(n) & h_{22}(n) \\ h_{31}(n) & h_{32}(n) \\ h_{41}(n) & h_{42}(n) \end{pmatrix}$$

Converting these matrix functions to transforms simply involves taking the transform of every element in each vector and matrix.

Matrix Convolutions

Consider the connection between a single input node i and a single output node j as though they were the only nodes of concern. If the system is linear, we should be able to calculate the output at node j by convolving the input at i with the system impulse response function. That is,

$$g_j(n \mid i) = \sum_{k=0}^{n} f_i(k) h_{ij}(n-k) \tag{7-45}$$

where $h_{ij}(n)$ is the impulse response at node j when the system is excited at node i. By taking transforms, we see that this reduces to the product

$$T(g_j(n \mid i)) = T(f_i(n)) T(h_{ij}(n)) \tag{7-46}$$

Now suppose that all M input nodes are simultaneously excited. The output at node j is then the sum of the outputs caused by each input acting separately. Symbolically we have

$$g_j(n) = \sum_{i=1}^{M} g_j(n \mid i) = \sum_{i=1}^{M} \sum_{k=0}^{n} f_i(k) h_{ij}(n-k) \tag{7-47}$$

By taking transforms of this expression, we find

$$T(g_j) = \sum_{i=1}^{M} T(f_i) T(h_{ij}) \tag{7-48}$$

for $j = 1, 2, \ldots, N$, where N is the total number of output nodes.

If we examine Eq. (7-48) carefully, we see that $T(g_j)$ is nothing more than the jth term in a vector resulting from the multiplication of a vector $\mathbf{T}(\mathbf{f})$ by the matrix $\mathbf{T}(\mathbf{H})$. The total output at all nodes in the system is then contained in the vector

$$\mathbf{T}(\mathbf{g}) = \mathbf{T}(\mathbf{f})\mathbf{T}(\mathbf{H}) \tag{7-49}$$

The jth element of $\mathbf{T}(\mathbf{g})$ is the transform of the output at node j from a linear system with a total of M input nodes and N output nodes.

Combined Systems

We are now in a position to analyze compound systems in which the simple subsystems may each have multiple input and output nodes. The algebra is quite similar to that used for single-node systems except that we will now be using vectors and matrices in the calculations. The differences involve the distinction between matrix algebra and ordinary algebra.

First, matrix multiplication is not commutative. This means that the order in which vectors and matrices are multiplied is important. It is *not* proper to reverse the terms in Eq. (7-49). Second, one must always keep the dimensions consistent. For example, a 5×4 transition matrix will require a 1×5 input vector and will yield a 1×4 output vector. Third, the matrix analog of division is matrix inversion. Finally, any matrix which is to be inverted must be square and nonsingular.

Considering the caveats of the previous paragraph, we can now rewrite the rules for series, parallel, and feedback connections in terms of transformed vectors and matrices as follows:

Series connections are replaced by

$$\mathbf{T}(\mathbf{H}) = \mathbf{T}(\mathbf{H}_1)\mathbf{T}(\mathbf{H}_2) \tag{7-50}$$

Parallel connections are replaced by

$$\mathbf{T}(\mathbf{H}) = \mathbf{T}(\mathbf{H}_1) + \mathbf{T}(\mathbf{H}_2) \tag{7-51}$$

Simple feedback loops are replaced by

$$\mathbf{T}(\mathbf{H}) = [\mathbf{I} - \mathbf{T}(\mathbf{H}_1)]^{-1} \tag{7-52}$$

where $\mathbf{I}$ represents the $M \times M$ identity matrix, $\mathbf{H}_1$ is an $M \times M$ element in the loop, and the negative exponent represents the matrix inversion operation.

Example Consider an international company which has separate domestic and foreign divisions. The domestic division handles three product lines and the foreign division handles two. Each division uses the sales history of their product lines to prepare forecasts of personnel, raw material tonnage, and operating capital requirements for the parent company. The parent company uses these estimates to prepare a cash flow estimate for the total corporation.

One might diagram these various information flows as shown in Figure 7-12. The domestic division operation will be a 3×3 matrix of transfer functions, the foreign division 2×3, and the parent company 3×1. The total system could be replaced by a single operator. The details of that reduction are left as an exercise for the reader.

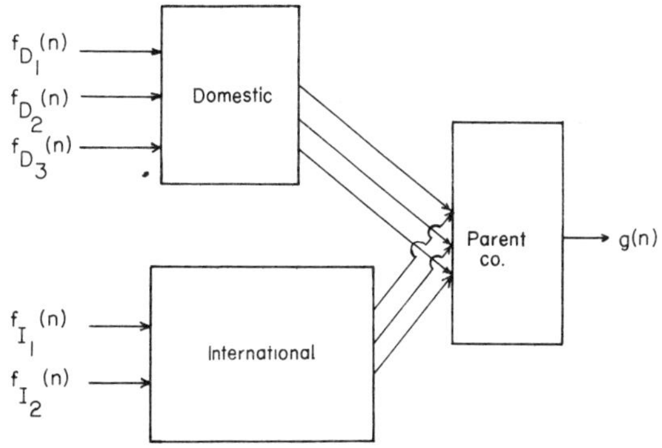

Figure 7-12 Multiple-node example.

Feedback in Markov Chains

The idea of multiple-node systems and feedback loops can be used to advantage in the analysis of Markov chains. Recall from the discussion in Chapter 5 that the basic recursion equation for a Markov chain can be written as

$$\mathbf{V}(n) = \mathbf{V}(n-1)\mathbf{P} \qquad \text{or} \qquad \mathbf{V}(n) = \mathbf{V}(0)\mathbf{P}^n \tag{7-53}$$

where $\mathbf{V}(n)$ is a vector whose elements are the state probabilities at time n and $\mathbf{P}$ is the $n \times n$ one-step transition matrix. The matrix $\mathbf{P}$ can be viewed as a system which takes an input vector at time $n - 1$ and produces the output vector $\mathbf{V}(n)$.

In using geometric transforms to analyze that system, we developed the transform

$$\mathbf{G}(\mathbf{P}^n) = (\mathbf{I} - z\mathbf{P})^{-1} \tag{7-54}$$

Comparing Eq. (7-54) with (7-52), we see that the iteration implied by the Markov chain calculation is equivalent to a linear system with a transfer function $z\mathbf{P}$ in a feedback loop. Since $\mathbf{P}$ is a constant in this model, the Markov chain acts as a feedback process with a delay of one time unit, as indicated by the presence of the variable z.

Howard makes extensive use of this relationship to analyze the behavior of a great many Markov processes [21, Chapter 3]. He also notes that any flow graph can be interpreted as a feedback matrix system, a fact which is fully exploited in the case of Markov chains. The details of the complete flow graph analysis of Markov chains are beyond the scope of this book. Interested readers are urged to consult Howard for this use of transforms and system diagrams.

Problems

1. A compound discrete system consisting of two simple elements hooked up in series has been constructed. The impulse response function for the first element has been found to be $h_1(n) = a^n(n + 1)$. When the second element was subjected to a pulse of magnitude 3, its response was found to be $h_2(n) = 3 \sin(-\frac{1}{2}\pi n)$.

(a) Find a transfer function for each element.

(b) Find the transfer function for a single element to replace the two in series.

(c) Use the above transfer function to find the response of the system to a driving function

$$f(n) = \begin{cases} 4 + 5(n - 2), & 2 \le n \le 6 \\ 0, & \text{elsewhere} \end{cases}$$

2. A discrete linear system has been subjected to a driving function $f(n)$ and has been observed to give a response $g(n)$ as follows:

n	$f(n)$	$g(n)$
0	1	0
1	3	1
2	2	8
3	1	−4
4	0	5
≥ 5	0	0

(a) Find the transfer function for the system.

(b) Find the impulse response function.

(c) Find the response of the system to a new input of the form

$$f_1(n) = \begin{cases} 5n + 3n^2, & 0 < n \le 4 \\ 0, & \text{elsewhere} \end{cases}$$

3. Consider a continuous system with the following structure:

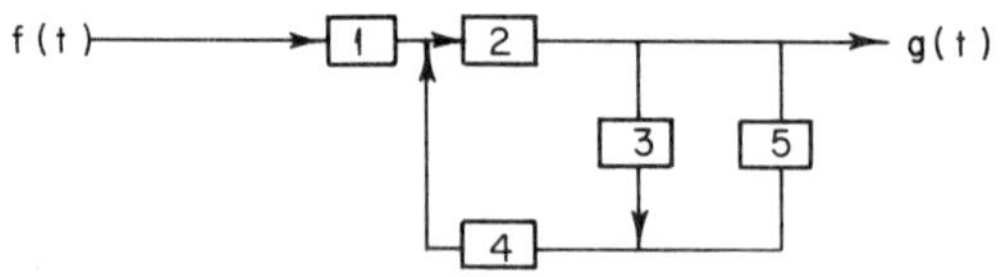

System 1: $$\frac{\mathscr{L}(\text{in})}{\mathscr{L}(\text{out})} = \frac{(s + 3)^2(s - 1)}{s}$$

System 2: $$\frac{\mathscr{L}(\text{out})}{\mathscr{L}(\text{in})} = e^{-2t}$$

System 3: $h(t) = 3\delta(t - 4)$

System 4: $\mathscr{L}(h(t)) = s^{-3}$

System 5: In a test configuration $f(t) = 3t$ gave

$$g(t) = t^2 - 2t.$$

(a) Find the impulse response function for every element.

(b) Find the transfer function for a single element to replace the compound system shown.

4. Select an operating system from your own experience which might profitably be analyzed by the techniques of this chapter.

(a) Describe the system verbally, including definitions of all information flows and major subsystems.

(b) Discuss the major difficulties you would expect to encounter in developing needed transfer functions. Suggest some methods for overcoming those difficulties.

(c) Draw a block diagram for this system.

(d) Reduce your system diagram to a single equivalent system element.

5. The impulse response function for a discrete linear system has been observed to be $h(n)$. A new input function $f(n)$ is to be presented to the system. Find the response of the system to this new input.

n	$h(n)$	$f(n)$
0	0	1
1	1	0
2	0	3
3	4	2
4	0	0

6. Consider the paper flow generated by requests for potential rabies treatment at the student health clinic. Because of his prior training, the clinic analyst is convinced that this must be a discrete linear system. He has coded a treatment request from a student bitten by a stray dog in order to judge the system response. Time is considered to be discrete, measured in days. Output is measured in number of documents generated by the request. The analyst notes the following:

(i) On the day of the request notices were mailed to the student's parents, the Dean, and the local humane society.

(ii) Two days later the student newspaper prepared a card on the case for future use in campus exposes. At the same time the editor made 3 copies for posting on the fraternity bulletin boards.

(iii) Three days after the request the city health inspector issued a written order for an interview with the student. That same day the Health Center gave the student an instruction booklet on what to do until the treatment starts.

(iv) On the fourth day a letter arrived from the student's parents demanding an audience with the health director.

(v) On the fifth day the student received a shot, a bill, and a disclaimer from the University denying all responsibility for further events. At that point the paper mill ceased to grind.

With the information now at hand, the analyst wants to measure the output of the system when the following number of requests are received:

Monday, 0, Tuesday, 1, Wednesday, 3, Thursday, 0, Friday, 15

Assume that the system is at rest on Monday. Develop the response function for this new input.

7. Consider the block diagram for an inventory system in Figure 7-10.

(a) Restructure the diagram using a transfer function for each simple element.

(b) Find a transfer function connecting sales and inventory.

(c) Find a transfer function connecting sales and production.

(d) Find the inventory and production levels generated for a 6-month period by a system with initial inventory of 5 vehicles and a sales pattern of $s(n) = 5 + 0.2n + \sin(\pi n/2)$. State all assumptions.

8. Consider the block diagram of Figure 7-12.

(a) Find the transfer function for a single multiple-node system which will replace the diagram shown.

(b) Discuss the difficulties of implementing a model of this type in the multicompany complex posed.

9. The campus and statehouse welfare collection agencies both receive a stream of contributions from their various classes of contributors. The campus group receives input of the form $p_1(n)$ from professors, $p_2(n)$ from students, and $p_3(n)$ from administrators, where n measures the number of elapsed days since the beginning of the quarter. The statehouse group receives $g_1(n)$ from civil service employees and $g_2(n)$ from politicians. Both the campus and the statehouse groups provide subsidies for a boys' camp and a girls' camp, which can be viewed as input nodes to an adolescent life system. The output from the adolescent system consists of dollar flows to clothing manufacturers $a_1(n)$, breweries $a_2(n)$, and universities $a_3(n)$.

(a) Construct a flow graph model of this system with inputs at various nodes represented by vectors of transforms and the transfer function by matrices. Develop an expression for the transform of the system output vector. Indicate the dimensions on all matrices.

(b) When a dollar is received from a civil service employee, the statehouse group donates \$0.10 to the boys' camp 5 days later and \$0.20 to the girls' camp 8 days later. A dollar from a politician leads to \$0.80 to the boys' camp in 2 days and a stream of $n(0.5^n)$ dollars for the girls' camp n days after the contribution. Find a transfer-function matrix (using geometric transforms) for the statehouse system.

(c) Calculate the total dollar amount for the girls' camp generated by a single \$1 contribution by a politician.

APPENDIX A

FINITE CALCULUS

In the process of modeling many man–machine and economic systems, the analyst is faced with functions that can only be defined for integer arguments or he may be faced with functions defined only in tabular fashion. In constructing inventory models for an automobile dealer, for example, the number of cars on hand can only sensibly be described as an integer quantity. Furthermore, data from this system are probably at best a table of inventory by day. Similarly, in examining the waiting line at a tool crib, the number of mechanics in line must obviously be an integer. Although it is true that in many cases reasonable approximations can be obtained by assuming that such quantities are continuous, our concern here is for those systems in which the units are so valuable or so few in number that their integrality must be maintained.

The techniques of finite differences also find application in solving complicated differential equations by numerical methods and in other branches of numerical analysis. Among the many areas employing these tools are numerical differentiation and integration, curve fitting, and interpolation. In a sense, finite differences are perfect examples of the engineering approach to problems. Data rarely come to the analyst in closed form such as sin $3t$. He must use numerical methods to analyze system input and response. We want to explore some of these techniques.

The purpose of this appendix is to introduce the reader to some of the basic concepts from the calculus of finite differences that we can apply to the analysis of physical systems. The strategy will be to demonstrate the close parallel between the operations in finite calculus and the more familiar procedures of differential and integral calculus. For an in-depth coverage of the topics that follow the reader should consult the texts by Miller [29] and Hildebrand [19].

Difference Equations

In the differential calculus the derivative of a function is defined in terms of the limit of a ratio as follows:

$$\frac{df(x)}{dx} = \lim_{h\to 0} \frac{f(x+h) - f(x)}{h} \tag{A-1}$$

where $f(x)$ is defined on some interval $[a, b]$ and x and $x + h$ are in $[a, b]$. If this limit exists, it is called the derivative of $f(x)$ at the point x. The symbols $f'(x)$ and $Df(x)$ are also used to represent this operation. In modeling linear systems, differential equations involving $(d/dx)f(x)$, $(d^2/dx^2)f(x)$, and higher order derivatives are formed. The task is then to find the function $f(x)$ satisfying the differential equation.

In difference calculus, we can consider equations in which the system is defined in terms of functions and changes in functions evaluated at particular points, say x, $x + k$, $x + j$, etc. Such expressions can be treated as difference equations in which infinitesimal increments in a function are replaced by larger quantities. One can define the first forward difference as the same basic ratio as a derivative but without the limiting argument. Let us define the difference quotient, sometimes referred to as the first divided difference, as

$$\Delta_h f(x) = \frac{f(x+h) - f(x)}{h} \tag{A-2}$$

Note that this is the same function that appears in Eq. (A-1). When one omits the divisor h, this quantity reduces to simply the "difference" of the function $f(x)$.

In addition to the forward difference, there is another operator ∇, which defines the first backward difference. By convention

$$\nabla_h f(x) = \frac{f(x) - f(x-h)}{h} \tag{A-3}$$

Having noted the existence of such an operator, the balance of our discussion will concentrate on the forward difference and its many parallels with the derivative.

To facilitate calculations, it is customary to assume that the values of x are equally spaced. The differences of the function are then based on sets of consecutive functional values. If the constant difference between consecutive values of x is h, we can define the general value of x for the tabular function as

$$x_k = x_0 + kh, \qquad k = \ldots, -2, -1, 0, 1, 2, \ldots \tag{A-4}$$

The corresponding functional value is then

$$f_k = f(x_0 + kh) \tag{A-5}$$

The first difference can then be written as a function of k as

$$\Delta f_k = f_{k+1} - f_k \tag{A-6}$$

If the constant difference between consecutive values is unity, i.e., $x = 0$, 1, 2, ..., Eq. (A-6) reduces to

$$\Delta f(x) = f(x+1) - f(x) \tag{A-7}$$

This first forward difference can also be viewed as the difference quotient of Eq. (A-2) in which the divisor is unity. In the calculations that follow, the increment h will be assumed to be unity unless otherwise stated.

Results obtained by manipulating this normalized equation can be converted for use in any problem with a constant increment h of arbitrary magnitude by the following argument. Let

$$r = x/h \tag{A-8}$$

and define a function g such that

$$g\left(\frac{x}{h}\right) = \frac{1}{h} f(x) \tag{A-9}$$

from which

$$\frac{1}{h} f(x+h) = g\left(\frac{x+h}{h}\right) = g(r+1)$$

It then follows that

$$\frac{f(x+h) - f(x)}{h} = g(r+1) - g(r) = \Delta g(r) \tag{A-10}$$

The implication of this result is that one can develop identities for the special case of a divided difference with $h = 1$ and use these results for manipulating functions with any uniform but otherwise arbitrary increment in the argument x.

General Formulas

The Δ operator is on the surface analogous to the differentiation operator d/dx. The next task is to see how far one can extend this analogy in the development of standard relationships routinely employed to shorten our calculations. Just as it is unnecessary to revert to the formation of the ratio and limit expression of Eq. (A-1) everytime one wants to obtain the derivative of a function, so is it unnecessary to revert to the explicit calculation of Eq. (A-7) to obtain the difference of a function. One uses instead a table of theorems and identities to facilitate the calculation.

To illustrate the close parallels, recall from basic calculus the theorems used to calculate such things as derivatives of sums, products, and quotients. For example, the derivative of a quotient of two functions can be expressed as a function of the numerator and denominator as

$$D\left(\frac{f(x)}{g(x)}\right) = \frac{[Df(x)]g(x) - [Dg(x)]f(x)}{[g(x)]^2} \tag{A-11}$$

To develop an analogous relationship for the difference calculation, let us begin with the basic definition

$$\Delta\frac{f(x)}{g(x)} = \frac{f(x+1)}{g(x+1)} - \frac{f(x)}{g(x)}$$

By placing the right side over a common denominator, we obtain

$$\Delta\frac{f(x)}{g(x)} = \frac{f(x+1)g(x) - g(x+1)f(x)}{g(x)g(x+1)}$$

In order to convert the numerator to an expression in differences of the f and g functions, let us add and subtract the common term $g(x)f(x)$:

$$\Delta\frac{f(x)}{g(x)} = \frac{f(x+1)g(x) - g(x+1)f(x) + g(x)f(x) - g(x)f(x)}{g(x)g(x+1)}$$

By regrouping terms in the numerator, we are led to

$$\begin{aligned}\Delta\frac{f(x)}{g(x)} &= \frac{g(x)[f(x+1) - f(x)] - f(x)[g(x+1) - g(x)]}{g(x)g(x+1)} \\ &= \frac{g(x)\,\Delta f(x) - f(x)\,\Delta g(x)}{g(x)g(x+1)} \qquad \text{(A-12)}\end{aligned}$$

Note the close resemblance of Eq. (A-12) to Eq. (A-11). The differences for the functions f and g play the same role as did the corresponding derivatives when taking the derivative of a quotient. The only dissimilarity occurs in the denominator of the difference of the quotient, where we now have a product of $g(x)g(x+1)$ in place of the square of the g-function found in the derivative.

Other relationships demonstrating a striking resemblance to the analogous differentiation procedure are equally easy to prove. For example, the derivative of the sum of two functions is treated as the sum of derivatives. That is,

$$D[f(x) + g(x)] = Df(x) + Dg(x) \tag{A-13}$$

In similar fashion we have

$$\begin{aligned}\Delta[f(x) + g(x)] &= [f(x+1) + g(x+1)] - [f(x) + g(x)] \\ &= [f(x+1) - f(x)] + [g(x+1) - g(x)] \\ &= \Delta f(x) + \Delta g(x)\end{aligned} \tag{A-14}$$

Other such analogies, some of which are tabulated in Table A-1, are left as exercises for the reader.

Higher Order Differences

Just as one is often concerned with higher order derivatives of functions in many applications, so may we be concerned with higher order differences. In differential calculus one defined recursively the nth derivative as

$$D^n f(x) = \frac{d^n}{dx^n} f(x) = \frac{d}{dx}\left[\frac{d^{n-1}}{dx^{n-1}} f(x)\right] \tag{A-15}$$

In like manner one can define the nth difference as

$$\Delta^n f(x) = \Delta[\Delta^{n-1} f(x)] \tag{A-16}$$

For example, the second difference of $f(x)$ is simply the difference of the first difference.

$$\begin{aligned}\Delta^2 f(x) &= \Delta[\Delta f(x)] = \Delta[f(x+1) - f(x)] \\ &= [f(x+2) - f(x+1)] - [f(x+1) - f(x)] \\ &= f(x+2) - 2f(x+1) + f(x)\end{aligned} \tag{A-17}$$

Continuing in this recursive fashion, one can develop a general expression for the nth difference as

$$\Delta^n f(x) = \sum_{i=0}^{n} (-1)^{n+i} \binom{n}{i} f(x+i) \tag{A-18}$$

Basic Identities

It is convenient to have a table of differences for commonly encountered functions that one can use in concert with the more general theorems involving sums, quotients, and products to solve particular problems. Again the identities developed in basic calculus have close parallels in the difference calculus.

For example, in basic calculus one learned early that

$$\frac{d}{dx}(e^x) = e^x \tag{A-19}$$

The analogous difference formula is

$$\Delta 2^x = 2^x \tag{A-20}$$

Proof for such statements proceeds in straightforward fashion by merely returning to the definition of the first forward difference, just as one once returned to the limiting arguments in calculus. That is,

$$\Delta 2^x = 2^{x+1} - 2^x = 2^x(2 - 1) = 2^x$$

Of course once this relationship has been satisfactorily proven it is stored in the proper table for our use, just as was the derivative of e^x. There is no need to "reinvent the wheel" every time one must find the difference of a function, just as there is no need when one seeks the derivative of a function.

Several identities for commonly encountered functions are contained in Table A-1. The proofs are left as exercises for the reader.

Factorial Expressions

The analogy between differences and derivatives becomes even more striking when we introduce the concept of factorial polynomials. It is usual to define x, n-factorial, as

$$x^{(n)} = x(x - 1) \cdots (x - n + 1) \tag{A-21}$$

and

$$x^{-(n)} = \frac{1}{(x + 1)(x + 2) \cdots (x + n)} \tag{A-22}$$

for integer arguments of n. However, both cases can be subsumed under the single expression

$$x^{(n)} = \frac{x!}{(x - n)!} \tag{A-23}$$

where n may be either positive or negative. For noninteger argument n, this expression can be written as

$$x^{(n)} = \frac{\Gamma(x+1)}{\Gamma(x-n+1)}, \qquad \text{where} \quad \Gamma(n) = \int_0^\infty t^{n-1}e^{-t}\,dt$$

The exponent within parentheses will be reserved in this discussion for factorial expressions. Those expressions containing exponents without parentheses will be the common power polynomials with which we usually work.

It is also possible to form factorial powers of functions of x. They are defined similarly to Eqs. (A-21) and (A-22) as

$$f(x)^{(n)} = f(x)f(x-1)\cdots f(x-n+1) \tag{A-24}$$

and

$$f(x)^{-(n)} = [f(x+1)f(x+2)\cdots f(x+n)]^{-1} \tag{A-25}$$

The only caveat here is that one should always perform the expansion before evaluating such functions for particular values of the variable x.

In general, the factorial expressions permit us to operate in the calculus of finite differences in much the same way that the power functions x^n are used ordinary calculus. For example, suppose that we seek the first difference of $x^{(n)}$. By applying the definitions, we obtain

$$\begin{aligned}
\Delta x^{(n)} &= (x+1)^{(n)} - x^{(n)} \\
&= [(x+1)(x)(x-1)\cdots(x-n+2)] \\
&\quad - [x(x-1)(x-2)\cdots(x-n+1)] \\
&= (x+1)[x(x-1)\cdots(x-n+2)] \\
&\quad - (x-n+1)[x(x-1)\cdots(x-n+2)] \\
&= [(x+1)-(x-n+1)]x^{(n-1)} \\
&= nx^{(n-1)}
\end{aligned} \tag{A-26}$$

If the exponent had been negative, it is a simple matter to show that

$$\Delta x^{-(n)} = -nx^{-(n+1)} \tag{A-27}$$

The resemblance to the derivative of a power expression is unmistakable. That is,

$$\frac{d}{dx}x^n = nx^{n-1} \tag{A-28}$$

When linear combinations of factorial expressions are used, we will refer to such expressions as factorial polynomials. An equation such as

$$q(x) = 5x^{(3)} + 2x^{-(2)} - 6x^{(1)}$$

is a factorial polynomial. If one wants the difference of q, direct application of Eqs. (A-26) and (A-27) yields

$$\Delta q(x) = 15x^{(2)} - 4x^{-(3)} - 6$$

Stirling Numbers

Due to the simplicity of this calculation, it is of value to be able to convert power expressions into factorial expressions and vice versa. The devices for accomplishing these conversions are the Stirling number relationships. What we seek are coefficients such that

$$x^{(n)} = \sum_{i=1}^{n} S_{i,n} x^i \tag{A-29}$$

and

$$x^n = \sum_{i=1}^{n} \mathscr{S}_{i,n} x^{(i)} \tag{A-30}$$

The terms $S_{i,n}$ are called Stirling numbers of the first kind and the terms $\mathscr{S}_{i,n}$ are called Stirling numbers of the second kind.

Low-order coefficients can be determined by simple algebraic expansion of the factorial expression. For example,

$$\begin{aligned} x^{(1)} &= x^1 \\ x^{(2)} &= x(x-1) = x^2 - x \\ x^{(3)} &= x(x-1)(x-2) = x^3 - 3x^2 + 2x \end{aligned} \tag{A-31}$$

from which it is obvious that

$$\begin{aligned} &S_{1,1} = 1 \\ &S_{1,2} = -1, \qquad S_{2,2} = 1 \\ &S_{1,3} = 2, \qquad S_{2,3} = -3, \qquad S_{3,3} = 1 \end{aligned} \tag{A-32}$$

Furthermore, by rearranging the terms in the relationships of (A-31), we see that

$$\begin{aligned} x^1 &= x^{(1)} \\ x^2 &= x^{(2)} + x^{(1)} \\ x^3 &= x^{(3)} + 3x^2 - 2x \\ &= x^{(3)} + 3(x^{(2)} + x^{(1)}) - 2x^{(1)} \end{aligned} \tag{A-33}$$

from which

$$\mathscr{S}_{1,1} = 1$$
$$\mathscr{S}_{1,2} = 1, \quad \mathscr{S}_{2,2} = 1 \tag{A-34}$$
$$\mathscr{S}_{1,3} = 1, \quad \mathscr{S}_{2,3} = 3, \quad \mathscr{S}_{3,3} = 1$$

Equations (A-33) and (A-34) are special cases. To develop coefficients for higher order polynomials, it is more convenient to use the general recursion formulas for Stirling numbers. In general

$$S_{i,n+1} = S_{i-1,n} - nS_{i,n} \tag{A-35}$$

and

$$\mathscr{S}_{i,n+1} = \mathscr{S}_{i-1,n} + i\mathscr{S}_{i,n} \tag{A-36}$$

For ease of computation it is convenient to form tables for particular values of the Stirling numbers much as one forms Pascal's triangle for binomial coefficients in basic algebra. The reader is urged to verify these equations and to prepare such tables of Stirling numbers for future computations.

Developing Factorial Polynomials

In many applications data will necessarily be presented in tabular form. In analyzing such data it is sometimes convenient to construct a difference table in which columns represent the various orders of differences for the function $f(x)$. For example, consider the following tabulation:

x	$f(x)$	$\Delta f(x)$	$\Delta^2 f(x)$	$\Delta^3 f(x)$	$\Delta^4 f(x)$
0	6	−1	12	12	0
1	5	11	24	12	0
2	16	35	36	12	
3	51	71	48		
4	122	119			
5	241				

(A-37)

There is a theorem in interpolation theory that states that if one has any set of $n + 1$ numbers, there exists a polynomial of degree not exceeding n that will pass through each of those numbers. In order to develop an expression for this polynomial, let us begin with a factorial polynomial, which we know from the last section can always be converted to a corresponding power expression by means of the Stirling numbers.

Suppose that the polynomial we seek can be written as

$$f(x) = a_0 + a_1 x^{(1)} + a_2 x^{(2)} + \cdots + a_5 x^{(5)} \tag{A-38}$$

By setting $x = 0$, we see that all terms other than a_0 in Eq. (A-38) will be zero since all contain an x. It therefore follows that

$$a_0 = f(0) \tag{A-39}$$

If we now take the first difference of Eq. (A-38), we obtain

$$\Delta f(x) = a_1 + 2a_2 x^{(1)} + 3a_3 x^{(2)} + \cdots + 5a_5 x^{(4)} \tag{A-40}$$

When this expression is evaluated at $x = 0$, all terms on the right except a_1 go to zero. Therefore

$$a_1 = \Delta f(0) \tag{A-41}$$

Continuing in like fashion for higher order differences, we find that

$$a_n = \frac{1}{n!}\Delta^n f(0) \tag{A-42}$$

Combining Eqs. (A-38) and (A-42), we obtain

$$f(x) = f(0) + \Delta f(0)x^{(1)} + \frac{1}{2}\Delta^2 f(0)x^{(2)} + \cdots + \frac{1}{5!}\Delta^5 f(0)x^{(5)} \tag{A-43}$$

for the fifth-degree factorial polynomial. This expression is referred to as Newton's formula or sometimes as Newton's forward interpolation formula because of its emphasis on the initial value of the function $f(0)$. The analogy with Maclaurin's expansion from ordinary calculus is apparent. For continuous functions we know that

$$f(x) = f(0) + f'(0)x + \frac{1}{2}f''(0)x^2 + \cdots + \frac{1}{n!}f^n(0)x^n \tag{A-44}$$

where the derivatives at $x = 0$ for the power expression play a role analogous to the corresponding order differences in Newton's formula.

Returning now to our particular difference table (A-37), a factorial polynomial for $f(x)$ can be written, using the numerical values for $\Delta^n f(0)$ in Eq. (A-43), as

$$f(x) = 6 + (-1)x^{(1)} + \tfrac{1}{2}(12)x^{(2)} + \tfrac{1}{6}(12)x^{(3)} = 6 - x^{(1)} + 6x^{(2)} + 2x^{(3)} \tag{A-45}$$

By expanding the factorial terms, we find the corresponding power polynomial to be

$$f(x) = 6 - x + 6[x(x - 1)] + 2[x(x - 1)(x - 2)] = 6 - 3x + 2x^3 \tag{A-46}$$

Sum Calculus

The Antidifference

Just as there is a strong analogy between difference calculus and differential calculus, so is there a strong analogy between the sum calculus and integral calculus. We can begin to explore this analogy by considering first a function that has been expressed as a series of factorial polynomials. What we now seek to do is to reverse the operations of Eqs. (A-26) and (A-27) to form a new function which has the given function as its first difference. For example, if $\Delta x^{(n)} = nx^{(n-1)}$, what is $\Delta^{-1}x^{(n)}$? This antidifference is analogous to the antiderivative of integral calculus.

In integral calculus one forms the antiderivative or indefinite integral of a polynomial function by

$$D^{-1}x^n = \int x^n \, dx = \frac{1}{n+1} x^{n+1} + c \tag{A-47}$$

where c is the constant of integration. The derivative of the function on the right will of course reverse the integration operation and return the function x^n. The analogous equation in sum calculus for factorial polynomials is

$$\Delta^{-1}x^{(n)} = \sum x^{(n)} = \frac{1}{n+1} x^{(n+1)} + k \tag{A-48}$$

where the periodic constant k plays the same role as the constant of integration. All we have done is to reverse the difference operation of Eq. (A-26).

The idea of an antidifference can obviously be carried forth to functions other than simple polynomials. If, for example, we ask the question, "What is the function that has a first difference of 2^x?," we need only to recall the relationship of Eq. (A-20). Since that equation relates

$$\Delta 2^x = 2^x$$

it follows that

$$\Delta^{-1}2^x = 2^x + k \tag{A-49}$$

Similarly, the antidifference of a^x can be developed by reversing the difference operation

$$\Delta a^x = a^x(a-1)$$

By taking the antidifference of both sides of this expression, and noting that

$$\Delta^{-1}\,\Delta f(x) = f(x)$$

we find that

$$a^x = \Delta^{-1} a^x(a - 1) + k$$

from which

$$\Delta^{-1} a^x = \frac{a^x}{a - 1} + k \qquad \text{(A-50)}$$

Equations for the antidifferences of other common functions, together with a number of general formulas analogous to integration operations, are listed in Table A-2 at the end of this appendix.

Sums of Series

The ability to determine antidifferences provides a convenient method to evaluate sums of series. To illustrate why this is true, let us return to the difference table representation of a function.

Consider any two consecutive columns of the difference table for the function $g(x)$, say columns n and $n + 1$:

$$\begin{array}{ll} \Delta^n g(0) & \Delta^{n+1} g(0) \\ \Delta^n g(1) & \Delta^{n+1} g(1) \\ \vdots & \vdots \\ \Delta^n g(r) & \Delta^{n+1} g(r) \\ \Delta^n g(r + 1) & \end{array} \qquad \text{(A-51)}$$

From the definition of a difference we know that

$$\Delta^{n+1} g(k) = \Delta^n g(k + 1) - \Delta^n g(k) \qquad \text{(A-52)}$$

It therefore follows that we could sum all of the terms in column $n + 1$ as follows:

$$\begin{aligned} \sum_{i=0}^{r} \Delta^{n+1} g(i) &= \sum_{i=0}^{r} [\Delta^n g(i + 1) - \Delta^n g(i)] \\ &= [\Delta^n g(1) - \Delta^n g(0)] + [\Delta^n g(2) - \Delta^n g(1)] \\ &\quad + \cdots + [\Delta^n g(r + 1) - \Delta^n g(r)] \end{aligned}$$

By canceling the common terms on the right, we find that this reduces to

$$\sum_{i=0}^{r} \Delta^{n+1} g(i) = \Delta^n g(r + 1) - \Delta^n g(0) \qquad \text{(A-53)}$$

But the nth difference of a function is obviously the antidifference of the

$(n + 1)$th difference. It therefore seems plausible to replace the term $\Delta^{n+1}g(i)$ by the function $f(i)$ and define $F(i)$ as the antidifference of $f(i)$. It then becomes clear that Eq. (A-53) is equivalent to

$$\sum_{i=0}^{r} f(i) = F(r + 1) - F(0) \tag{A-54}$$

This is a special case of the fundamental theorem of sum calculus. In general if we want to find the sum of a series $f(x)$, where x covers the interval $[a, b]$, we can develop a closed form expression by evaluating the antidifference or indefinite sum at the indicated limits. That is,

$$\sum_{x=a}^{b} f(x) = \sum f(x)\Big|_{a}^{b+1} = F(b + 1) - F(a) \tag{A-55}$$

The striking resemblance to the fundamental theorem of integral calculus is made apparent by recalling

$$\int_{a}^{b} f(x)\, dx = \int f(x)\, dx \Big|_{a}^{b} = F(b) - F(a) \tag{A-56}$$

Once a table of indefinite sums has been developed, it can be employed to evaluate any number of series in a manner analogous to the way one uses a table of integrals. This technique will be especially helpful in our later work with discrete transforms. There we will encounter sums of the type

$$G(z) = \sum_{i=0}^{n} z^i f(i) \tag{A-57}$$

where $f(i)$ is a known function. Facility with the sum calculus will often enable us to find closed form expressions for such transforms with a minimum amount of calculation.

Infinite Sums

In integral calculus one is often called upon to evaluate integrals in which the limits of integration may be infinite. The method of solution for such improper integrals is to treat them as though they had finite limits of integration, then take the limit as the limits of integration approach infinity. When such limits do not exist, the improper integral diverges and has no value. For example,

$$\int_{0}^{\infty} f(t)\, dt = \lim_{h\to\infty} \int_{0}^{h} f(t)\, dt \tag{A-58}$$

An analogous situation arises in sum calculus with the infinite series. The solution strategy is exactly the same one employed for infinite integrals. Namely, treat the sum as though it had finite limits, then take the limit of the result as the finite terms go to infinity. That is,

$$\sum_{x=a}^{\infty} f(x) = \lim_{b\to\infty} \sum_{x=a}^{b} f(x) \tag{A-59}$$

Examples To illustrate the above techniques, let us evaluate the well-known sums

$$\sum_{t=1}^{n} t \qquad \text{and} \qquad \sum_{t=1}^{n} t^2$$

As usual, it is easier to work with factorial polynomials than with power polynomials in the sum calculus. Therefore, converting to factorial terms, we have

$$\sum_{t=1}^{n} t = \sum_{t=1}^{n} t^{(1)}$$

which is readily evaluated as

$$\sum_{t=1}^{n} t^{(1)} = \tfrac{1}{2}t^{(2)}\Big|_{1}^{n+1} = \tfrac{1}{2}[(n+1)(n) - 0] = \tfrac{1}{2}n(n+1) \tag{A-60}$$

In similar fashion the sum of squares is calculated from

$$\sum_{t=1}^{n} t^2 = \sum_{t=1}^{n} (t^{(2)} + t^{(1)}) = \tfrac{1}{3}t^{(3)} + \tfrac{1}{2}t^{(2)}\Big|_{1}^{n+1} = \tfrac{1}{6}n(n+1)(2n+1) \tag{A-61}$$

Now consider the infinite series

$$\sum_{x=1}^{\infty} f(x) = \frac{1}{6} + \frac{1}{24} + \frac{1}{60} + \cdots$$

Rewriting the denominator as a product of terms, we obtain

$$\sum_{x=1}^{\infty} f(x) = \frac{1}{1\cdot 2\cdot 3} + \frac{1}{2\cdot 3\cdot 4} + \frac{1}{3\cdot 4\cdot 5} + \cdots$$

which can be recognized as

$$\sum_{x=1}^{\infty} f(x) = \sum_{x=1}^{\infty} \frac{1}{x(x+1)(x+2)} = \sum_{x=0}^{\infty} x^{-(3)}$$

Applying the basic identities, we find

$$\sum_{x=0}^{\infty} x^{-(3)} = \lim_{h\to\infty} \sum_{x=0}^{h} x^{-(3)} = \lim_{h\to\infty} -\frac{x^{-(2)}}{2}\Bigg|_{0}^{h+1}$$

$$= \lim_{h\to\infty} -\frac{1}{2(x+1)(x+2)}\Bigg|_{0}^{h+1} = \lim_{h\to\infty}\left[-\frac{1}{2(h+2)(h+3)} + \frac{1}{2(2)}\right] = \frac{1}{4}$$

As a final illustration, consider the geometric sequence

$$g = \sum_{i=0}^{\infty} r^i$$

From our previous work with sums we know that

$$\sum a^x = \frac{a^x}{a-1}$$

Therefore

$$\sum_{i=0}^{\infty} r^i = \lim_{h\to\infty} \frac{r^i}{r-1}\Bigg|_{i=0}^{h+1} = \lim_{h\to\infty}\left(\frac{r^{h+1}}{r-1} - \frac{1}{r-1}\right)$$

For the special case of $r < 1$, this reduces to the well-known result

$$\sum_{i=0}^{\infty} r^i = \frac{1}{1-r}$$

Useful Relationships

In order to expedite our calculations involving finite calculus, a number of useful relationships for differences and sums are listed in Tables A-1 and A-2. Some of these were discussed in the text but most are presented without proof. The proofs are for the most part straightforward. They are suggested as problem exercises to supplement the reader's understanding of this material.

Before leaving this subject, it is well to note summation by parts, which is perhaps the most important identity in sum calculus from the point of view of expanding tables with a limited number of entries to more complicated functions. Summation by parts is closely analogous to integration by parts from integral calculus. The relationship is

$$\sum f(n)\,\Delta g(n) = f(n)g(n) - \sum g(n+1)\,\Delta f(n) \tag{A-62}$$

For example, to evaluate $\sum nk^n$, let

$$f(n) = n \qquad \text{and} \qquad \Delta g(n) = k^n$$

from which $\Delta f(n) = 1$ and $g(n) = \sum k^n = k^n/(k-1)$. Then

$$\sum nk^n = \frac{nk^n}{k-1} - \sum \frac{k^{n+1}}{k-1}(1) = \frac{nk^n}{k-1} - \frac{k}{k-1}\left(\frac{k^n}{k-1}\right)$$
$$= \frac{k^n[n(k-1)-k]}{(k-1)^2}$$

We will operate in true engineering fashion from this point on by using the identities in these tables in the same manner as we use handbooks containing tables of integrals. Those readers interested in the mathematical subtleties or who wish to become true experts in the area are urged to consult one of the references listed at the end of the book.

Table A-1

Table of Differences

	$f(x)$	$\Delta f(x) = f(x+1) - f(x)$
1.	$g(x) + h(x)$	$\Delta g(x) + \Delta h(x)$
2.	$\Delta^{n-1} g(x)$	$\Delta^n g(x) = \sum_{i=0}^{n} (-1)^{n+i} \binom{n}{i} g(x+i)$
3.	$ag(x)$, a = constant	$a\,\Delta g(x)$
4.	$\dfrac{g(x)}{h(x)}$	$\dfrac{h(x)\,\Delta g(x) - g(x)\,\Delta h(x)}{h(x)h(x+1)}$
5.	$g(x)h(x)$	$g(x)\,\Delta h(x) + h(x+1)\,\Delta g(x)$ $= h(x)\,\Delta g(x) + g(x+1)\,\Delta h(x)$
6.	$\sum g(x) + k$	$g(x)$
7.	c, c = constant	0
8.	$x^{(n)}$	$nx^{(n-1)}$
9.	$(p+qx)^{(n)}$	$qn(p+qx)^{(n-1)}$
10.	2^x	2^x
11.	b^x	$b^x(b-1)$
12.	e^{ax}	$e^{ax}(e^a - 1)$
13.	$\ln(x)$	$\ln\left(1 + \frac{1}{x}\right)$
14.	$\log_b(x)$	$\log_b\left(1 + \frac{1}{x}\right)$
15.	$\sin(ax)$	$2(\sin\frac{1}{2}a)\cos[a(x+\frac{1}{2})]$
16.	$\cos(ax)$	$-2(\sin\frac{1}{2}a)\sin[a(x+\frac{1}{2})]$

Table A-2

Table of Indefinite Sums[a]

	$f(x)$	$\sum f(x)$
1.	$g(x) + h(x)$	$\sum g(x) + \sum h(x)$
2.	$ag(x)$, $a =$ constant	$a \sum g(x)$
3.	$g(x)\, \Delta h(x)$, summation by parts	$g(x)h(x) - \sum h(x+1)\, \Delta g(x)$
4.	$\Delta g(x)$	$g(x)$
5.	c	cx
6.	2^x	2^x
7.	b^x	$\dfrac{b^x}{b-1}$, $b \neq 1$
8.	$x^{(n)}$	$\dfrac{x^{(n+1)}}{n+1}$, $n \neq -1$
9.	e^{ax}	$\dfrac{e^{ax}}{e^a - 1}$
10.	$\sin(ax)$	$\dfrac{-\cos[a(x - \frac{1}{2})]}{2 \sin(\frac{1}{2}a)}$, $a \neq 2n\pi$
11.	$\cos(ax)$	$\dfrac{\sin[a(x - \frac{1}{2})]}{2 \sin(\frac{1}{2}a)}$, $a \neq 2n\pi$
12.	$(p + qx)^{(n)}$, p and q constant	$\dfrac{(p+qx)^{(n+1)}}{q(n+1)}$, $n \neq -1$
13.	nk^n	$\dfrac{k^n[n(k-1) - k]}{(k-1)^2}$

[a] The periodic constant of summation has been eliminated but should be added when using these identities.

APPENDIX B

TABLE OF LAPLACE TRANSFORMS

$f(t)$	$\mathscr{L}(f(t)) = F(s) = \int_0^\infty f(t)e^{-st}\,dt$
1. $ag(t)$	$a\mathscr{L}(g(t))$
2. $ag(t) + bh(t)$	$a\mathscr{L}(g(t)) + b\mathscr{L}(h(t))$
3. $\dfrac{dg(t)}{dt}$	$s\mathscr{L}(g(t)) - g(0)$
4. $\dfrac{d^n g(t)}{dt^n}$	$s^n\mathscr{L}(g(t)) - \sum_{i=0}^{n-1} s^{n-1-i} \left.\dfrac{d^i g(t)}{dt^i}\right\|_{t=0}$
5. $e^{at}g(t)$	$G(s - a)$
6. $g(t - a)U(t - a)$ where $U(t) = \begin{cases} 1, & t \geq 0 \\ 0, & \text{elsewhere} \end{cases}$	$e^{-as}\mathscr{L}(g(t))$
7. $\int_0^t g(r)\,dr$	$\dfrac{1}{s}\mathscr{L}(g(t))$

	$f(t)$	$\mathscr{L}(f(t)) = F(s) = \int_0^\infty f(t)e^{-st}\,dt$
8.	$\int_0^t g(r)h(t-r)\,dr$	$G(s)H(s)$
9.	$tg(t)$	$-\frac{d}{ds}G(s)$
10.	$t^n g(t)$	$(-1)^n \frac{d^n}{ds^n} G(s)$
11.	$\frac{1}{t}g(t)$	$\int_s^\infty G(s)\,ds$
12.	$g(at),\quad a > 0$	$\frac{1}{a}G\left(\frac{s}{a}\right)$
13.	$\lim_{t\to 0} f(t)$	$\lim_{s\to\infty} sF(s)$
14.	$\lim_{t\to\infty} f(t)$	$\lim_{s\to 0} sF(s)$ [valid for $F(s)$ with poles, excluding the origin, that lie in the left half of the s-plane]
15.	$g(t)$ where $g(t+a) = g(t)$	$\frac{\int_0^a e^{-st}g(t)\,dt}{1-e^{-as}}$
16.	$\delta(t)$ where $\int_{-\infty}^{\infty} \delta(t)\,dt = 1$ and $\delta(t) = 0,\quad t \neq 0$	1
17.	$\delta(t-a)$	e^{-as}
18.	k	$\frac{k}{s}$
19.	$t^n,\quad n = 1, 2, 3, \ldots$	$\frac{n!}{s^{n+1}}$
20.	$t^a,\quad a > 0$	$\frac{\Gamma(a+1)}{s^{a+1}}$
21.	e^{at}	$\frac{1}{s-a}$
22.	te^{at}	$\frac{1}{(s-a)^2}$

	$f(t)$	$\mathscr{L}(f(t)) = F(s) = \int_0^\infty f(t)e^{-st}\,dt$
23.	$\dfrac{e^{at}t^{n-1}}{\Gamma(n)}$	$\dfrac{1}{(s-a)^n}$
24.	$\dfrac{e^{a_1 t} - e^{a_2 t}}{a_1 - a_2}$	$\dfrac{1}{(s-a_1)(s-a_2)}$
25.	$\sin(at)$	$\dfrac{a}{s^2 + a^2}$
26.	$\cos(at)$	$\dfrac{s}{s^2 + a^2}$
27.	$\sinh(at)$	$\dfrac{a}{s^2 - a^2}$
28.	$\cosh(at)$	$\dfrac{s}{s^2 - a^2}$
29.	$(b-a)^{-1}, \quad a \le t \le b$	$\dfrac{e^{-as} - e^{-bs}}{s(b-a)}$
30.	$U(t-a) = \begin{cases} 0, & t < a \\ 1, & t > a \end{cases}$	$\dfrac{e^{-as}}{s}$
31.	ae^{-at}	$\dfrac{a}{s+a}$
32.	$\dfrac{t^{a-1}e^{-t/b}}{\Gamma(a)b^a}$	$\left(\dfrac{b^{-1}}{s+b^{-1}}\right)^a$
33.	$[g_{ij}(t)]$, a matrix	$[G_{ij}(s)]$

APPENDIX C

TABLE OF GEOMETRIC AND *z*-TRANSFORMS

Function $f(n)$	Geometric transform $G(f(n)) = G_n(z) = \sum_{n=0}^{\infty} f(n)z^n$	z-Transform $Z(f(n)) = Z_n(z) = \sum_{n=0}^{\infty} f(n)z^{-n}$
1. $k_1 g(n) + k_2 h(n)$	$k_1 G(f(n)) + k_2 G(h(n))$	$k_1 Z(g(n)) + k_2 Z(h(n))$
2. $g(n+1)$	$\dfrac{G(g(n)) - g(0)}{z}$	$z(Z(g(n)) - g(0))$
3. $g(n+r)$	$z^{-r}\left(G(g(n)) - \sum_{i=0}^{r-1} g(i)z^i\right)$	$z^r\left(Z(g(n)) - \sum_{i=0}^{r-1} g(i)z^{-i}\right)$
4. $g(n-r)U(n-r)$, where $U(n-r) = \begin{cases} 1, & n > r \\ 0, & \text{elsewhere} \end{cases}$	$z^r G(g(n))$	$z^{-r} Z(g(n))$
5. $\delta(n-r) = \begin{cases} 1, & n = r \\ 0, & n \neq r \end{cases}$	z^r	z^{-r}

Function $f(n)$	Geometric transform $G(f(n)) = G_n(z) = \sum_{n=0}^{\infty} f(n)z^n$	z-Transform $Z(f(n)) = Z_n(z) = \sum_{n=0}^{\infty} f(n)z^{-n}$
6. $\Delta g(n)$	$(z^{-1} - 1)G(g(n)) - z^{-1}g(0)$	$(z - 1)Z(g(n)) - zg(0)$
7. $\sum_{i=0}^{n} g(i)h(n-i)$	$G(g(n))G(h(n))$	$Z(g(n))Z(h(n))$
8. $k^{rn}g(n)$	$G(g(n))\Big\vert_{z=k^r z}$	$Z(g(n))\Big\vert_{z=z/k^r}$
9. 1	$\frac{1}{1-z}$	$\frac{z}{z-1}$
10. k	$\frac{k}{1-z}$	$\frac{kz}{z-1}$
11. k^n	$\frac{1}{1-kz}$	$\frac{z}{z-k}$
12. n	$\frac{z}{(1-z)^2}$	$\frac{z}{(z-1)^2}$
13. n^r	$\left(z\frac{d}{dz}\right)^r \frac{1}{1-z}$	$\left(-z\frac{d}{dz}\right)^r \left(\frac{z}{z-1}\right)$
14. $n^{(r)}$	$\frac{r!\,z^r}{(1-z)^{r+1}}$	$\frac{r!\,z}{(z-1)^{r+1}}$
15. $\frac{k^n}{n!}$	e^{kz}	$e^{k/z}$
16. nk^n	$\frac{kz}{(1-kz)^2}$	$\frac{kz}{(z-k)^2}$
17. $n^r k^n$	$\left(z\frac{d}{dz}\right)^r \left(\frac{1}{1-kz}\right)$	$\left(-z\frac{d}{dz}\right)^r \left(\frac{z}{z-k}\right)$
18. $\frac{k^n}{n}$	$-\ln(1-kz)$	$\ln z - \ln(z-k)$
19. $\binom{n+r}{r}k^n$	$(1-kz)^{-(r+1)}$	$\frac{z^{r+1}}{(z-k)^{r+1}}$
20. $\sin(kn)$	$\frac{z \sin k}{z^2 + 1 - 2z\cos k}$	$\frac{z \sin k}{z^2 + 1 - 2z\cos k}$
21. $\cos(kn)$	$\frac{1 - z\cos k}{z^2 + 1 - 2z\cos k}$	$\frac{z(z - \cos k)}{z^2 + 1 - 2z\cos k}$

	Function $f(n)$	Geometric transform $G(f(n)) = G_n(z) = \sum_{n=0}^{\infty} f(n)z^n$	z-Transform $Z(f(n)) = Z_n(z) = \sum_{n=0}^{\infty} f(n)z^{-n}$
22.	$\binom{r}{n} p^n q^{r-n}$	$(pz + q)^r$	$\dfrac{(qz + p)^r}{z^r}$
23.	$\dfrac{a^n e^{-a}}{n!}$	$e^{-a(1-z)}$	$e^{-(a/z)(z-1)}$
24.	$p(1 - p)^n$	$\dfrac{p}{1 - (1 - p)z}$	$\dfrac{pz}{z - 1 + p}$
25.	$\lim_{n \to 0} g(n)$	$G_n(z)\Big\vert_{z=0}$	$Z_n(z)\Big\vert_{z=\infty}$
26.	$\lim_{n \to \infty} g(n)$	$\lim_{z \to 1} (1 - z)G(g(n))$ [Provided all poles of $G(g(n))$ lie outside the unit circle]	$\lim_{z \to 1} \left(\dfrac{z-1}{z}\right) Z(g(n))$ [Provided all poles of $Z(g(n))$ lie inside the unit circle]
27.	$[g_{ij}(n)]$, a matrix	$[G(g_{ij}(n)]$	$[Z(g_{ij}(n))]$

REFERENCES AND SUGGESTED READING

1. Aseltine, J. A., *Transform Method in Linear System Analysis*. New York: McGraw-Hill, 1958.
2. Beightler, C. S., Mitten, L. G., and Nemhauser, G. L., "A Short Table of Z-Transforms and Generating Functions," *Operations Research* **9** (4) (July–August 1961), 574–578.
3. Bellman, R. E., Kalaba, R. E., and Lockett, J., *Numerical Inversion of the Laplace Transform*. New York: Amer. Elsevier, 1966.
4. Churchill, R. V., *Operational Mathematics*, 3rd ed. New York: McGraw-Hill, 1972.
5. Cooper, G. R., and McGillem, C. D., *Probabilistic Methods of Signal and System Analysis*. New York: Holt, 1971.
6. Cooper, R. B., *Introduction to Queueing Theory*. New York: Macmillan, 1972.
7. Cox, D. R., and Miller, H. D., *The Theory of Stochastic Processes*. New York: Wiley, 1968.
8. Craig, E. J., *Laplace and Fourier Transforms for Electrical Engineers*. New York: Holt, 1964.
9. Doetsch, G., *Guide to the Applications of the Laplace and Z-Transforms*. Princeton, New Jersey: Van Nostrand-Reinhold, 1971.
10. Donahue, James D., *Products and Quotients of Random Variables and Their Applications*. Wright-Patterson Air Force Base, Ohio: Aerospace Research Laboratories, ARL 64-115, Office of Technical Services AD603667, 1964.
11. Drake, A. W., *Fundamentals of Applied Probability Theory*. New York: McGraw-Hill, 1967.
12. Epstein, B., "Some Applications of the Mellin Transform in Statistics," *Annals of Mathematical Statistics* **19** (1948), 370–379.
13. Feller, W., *An Introduction to Probability Theory and Its Applications, Vol. I*, 3rd ed. New York: Wiley, 1971.

14. Fox, C., "Some Applications of Mellin Transforms to the Theory of Bivariate Statistical Distributions," *Cambridge Philosophical Society Proceedings* **53** (1957), 620–628.
15. Giffin, W. C., *Introduction to Operations Engineering*. Homewood, Illinois: Richard D. Irwin, 1971.
16. Giffin, W. C., "The Use of Generating Functions in Determining the Characteristics of Lead Time Demand," *Journal of Industrial Engineering* **18** (3) (March 1967), 237–238.
17. Hahn, G. J., and Shapiro, S. S., *Statistical Models in Engineering* New York: Wiley, 1967.
18. Haugen, E. B., *Probabilistic Approaches to Design*. New York: Wiley, 1968.
19. Hildebrand, F. B., *Finite Difference Equations and Simulations*. Englewood Cliffs, New Jersey: Prentice-Hall, 1968.
20. Hogg, R. V., and Craig, A. T., *Introduction to Mathematical Statistics*. New York: Macmillan, 1970.
21. Howard, R. A., *Dynamic Probabilistic Systems, Vol. I: Markov Models*. New York: Wiley, 1971.
22. Jordan, C., *Calculus of Finite Differences*, 2nd ed., Bronx, New York: Chelsea, 1947.
23. Kendall, D. G., "Some Problems in the Theory of Queues," *Journal of the Royal Statistical Society B* **13** (1951), 151–158.
24. Kendall, D. G., "Stochastic Processes Occurring in the Theory of Queues and Their Analysis by Means of the Imbedded Markov Chain," *The Annals of Mathematical Statistics* **24** (1953), 338–54.
25. Koopman, B. O., "Air-Terminal Queues Under Time-Dependent Conditions," *Operations Res.* **20** (3) (November–December 1972), 1089–1114.
26. Kotlarski, I., "On Random Variables Whose Quotient Follows the Cauchy Law," *Colloquium Mathematicum* **8** (1960), 277–283.
27. Kreyszig, E., *Advanced Engineering Mathematics*. 3rd ed., New York: Wiley, 1972.
28. Laning, J. H., and Battin, R. H., *Random Processes in Automatic Control*. New York: McGraw-Hill, 1956.
29. Miller, K. S., *An Introduction to the Calculus of Finite Differences and Difference Equations*. New York: Holt, 1960.
30. Milne-Thomson, L. M., *The Calculus of Finite Differences*. New York: St. Martin's, 1951.
31. Saaty, T. L., *Elements of Queueing Theory with Applications*. New York: McGraw-Hill, 1961.
32. Schmidt, J. W., *Mathematical Foundations For Management Science and Systems Analysis*. New York: Academic, 1974.
33. Van Landingham, H. F., and Shariq, S. A., "Numerical Evaluation of Compound Poisson Distributions," *AIIE Transactions* **6** (1) (March 1974), pp. 50–54.
34. Winkler, R. L., Roodman, G. M., and Britney, R., "The Determination of Partial Moments," *Management Science* **19** (3) (November 1972), 290–296.
35. Wylie, C. R., Jr., *Advanced Engineering Mathematics*, 2nd ed. New York: McGraw-Hill, 1960.

INDEX

Q

R

S

A 5
B 6
C 7
D 8
E 9
F 0
G 1
H 2
I 3
J 4